Orchestrating Human-Centered Design

Guy André Boy

Orchestrating Human-Centered Design

Springer

Dr. Guy André Boy, Ph. D.
Human-Centered Design Institute
Florida Institute of Technology
Melbourne, FL
USA

NASA Kennedy Space Center
Orlando, FL
USA

Florida Institute for Human and Machine Cognition
Pensacola, FL
USA

ISBN 978-1-4471-6196-7   ISBN 978-1-4471-4339-0 (eBook)
DOI 10.1007/978-1-4471-4339-0
Springer London Dordrecht Heidelberg New York

Cover Image: Jeremy Boy (myjyby@gmail.com)

Springer is part of Springer Science+Business Media (www.springer.com)

Preface

I am at Charles de Gaulle airport, boarding to Toulouse in an Airbus A320, an aircraft that I know well. I look forward to seeing my French Southwest again. Coming from Florida where I live now, I am already dreaming about a well-cooked "pintade aux cèpes".[1] In my home region in France, food is essential not only for the pleasure of taste, but also for social reasons. Lots of business happens around a good table with excellent libations. Human-centered activities do not mix with fast operations, except when speed can bring pleasure. Fast food will never provide support to good and deep relationships as a well-cooked dinner done by people who love you and love what they do! Human-centered design follows a similar track; it takes expertise, experience, time and dedication to "cook" a good system. We will see later in this book that maturity is one of the main concepts that drive good design and development practice.

During the twentieth century, industry developed and produced massive amounts of technology, which significantly impacted every day human and machine activities from the Industrial Revolution to present. We have started to realize that engineering is not only a matter of technology, it must involve people and organizations. Current systems engineering is matter of dry processes and does not include people and organizations in meaningful ways. In the beginning of the twenty-first century, we fell into a paradox that suggests we need a different breed of people who articulate technology and organizations. Why? This phenomenon exists because technology and organizations are almost exclusively managed by finance people.

Finance is a tool as many other tools. Economies are constantly designed and adapted to the evolution of our societies. However, we should not forget that economies must support ecology, in their primary sense, i.e., "the care of the house." When I observe the evolution of our world around us, I conclude that now it works the other way around, i.e., ecological systems support economical systems! In particular, this means that people have to adapt to the structures and functions of our finance-driven society. This is not human-centered at all! In several industrial sectors, such as telecommunication and automobile, we started to observe many people, employees, literally killing themselves because they could no longer adapt anymore

---

[1] Guinea fowl with porcini mushrooms.

to the finance-driven constraints.[2] Therefore, human-centered design should not be limited to local design and use of artifacts. It should take into account long-term issues from the beginning. Interestingly, artificial complexity of our world today results from this lack of long-term planning.

In this book, I would like to share with you a humanistic view of technology and organizations. I would also like to share with you my enthusiasm with human-centered design. This is what I have been trying to do for the last 35 years or so, professionally; I thought it was a good time to provide a novel rationalization of these views. I strongly believe that scientists and engineers, as well as the public, should be more involved in the technological and societal changes that will happen during the next decades. Human kind always evolved by adaptation since the beginning of humanity, but today socio-technical evolution is much faster than it had ever been. We need to use and invent conceptual tools to understand this evolution and master it.

I am optimistic that human-centered design will be a vital engineering genre of the future where hard and soft sciences, well-cooked with a subtle dose of Art, will contribute to improving our lives. Over time, people have used their intelligence to design and develop artifacts that in turn supported them in improving their knowledge, knowhow and well-being. The Internet is a good example.

This book would not have existed without the support of many people who I would like to recognize here. First, I would like to thank my fellow colleagues of the Air and Space Academy who supported the organization of the 2008 Risk Taking conference for the motivating discussions we had on this subject and related topics. A booklet[3] was produced and I consider that this book is a follow-up.

The ACM[4] community helped me shape my understanding of the evolution of human factors in computing systems. I was very much inspired by the work of Don Norman, Terry Winograd and Jonathan Grudin. I thank them for the many discussions we had on human-computer interaction and computer-supported cooperative work. Thanks also to the HCI-Aero community who I follow and support since 1986.

During the last few years, I have had a constant assistance from Ludovic Loine, a nuclear engineer and HFE manager at AREVA, who not only provided support to my PhD students but also believed in the work I am doing. Discussions I had with Anabela Simoes, professor of ergonomics and specialist of aging across active life, provided me with more confidence in the direction I took; sorry Anabela if I did not develop aging issues. I owe a lot to my colleagues at Florida Institute of Technology and Kennedy Space Center for creative and rigorous debates on human-centered design.

Many thanks to colleagues in various parts of the world who helped me giving lectures and/or shaping my arguments during the early days of the FIT Ph.D. program

---

[2] Alemanno, S. P., & Cabedoche, B. (2011). Suicide as the ultimate response to the effects of globalisation? France TéLécom, psychosocial risks, and communicational implementation of the global workplace. *Intercultural Communication Studies, XX,* 2.

[3] Boy, G. A., & Brachet, G. (2008). Risk taking. dossier of the air and space academy. Toulouse: ASA.

[4] Association for Computing Machinery.

in human-centered design, including Thierry Bellet, Jeffrey Bradshaw, Marco Carvalho, Michael Conroy, Ondrej Doule, Ken Ford, Kerry Gilger, Gudela Grote, John Hansman, Erik Hollnagel, Daniel Krob, Semen Koksal, Paul Krois, Amy Pritchett, Morten Lind, Rebecca Mazzone, Patrick Millot, Jean Pinet, Myroslav Sparavalo, Etienne Tarnowsky, and George Wears.

My Ph.D. students contributed directly and indirectly to the content of this book. First, they forced me to rationalize human-centered design the way it is presented here, because I needed to create and articulate lectures I gave at FIT during the last 3 years. More specifically, I owe a big thank you to Sharon Chinoy and Nikki Hoier who patiently reviewed the whole draft. Rhonda Lyons, Don Platt, Kara Schmitt, and Lucas Stephane contributed to improve the quality of this book.

Finally, my children encouraged me to write on human-centered design and include more artistic perspectives. Perrine and Jeremy, I am proud of you and your current endeavors in design.

Melbourne, Florida   Guy André Boy
May 2012

Contents

Acronyms

AAAS	American Association for the Advancement of Science
ABS	Anti-lock Braking System
ACM	Association for Computing Machinery
ADS-B	Automatic Dependent Surveillance-Broadcast
AIM	Advanced Interaction Media
ARPANET	Advanced Research Projects Agency Network
ARTCC	Air Route Traffic Control Center
ASAS	Airborne Separation Assistance System
ASN	French Nuclear Safety Authority (Autorité de Sûreté Nucléaire)
ASR	Automatic Slip Regulation, Acceleration Slip Regulation
ASRS	Aviation Safety Reporting System
ATC	Air Traffic Control
ATCO	ATC controller
ATM	Air Traffic Management
AUTOS	Artifact, User, Task, Organization and Situation (framework)
AVZ	Appropriate Volumetric Zone
BITE	Built-In Test Equipments
CAD	Computer-Aided Design
CDTI	Cockpit Display of Traffic Information
CFR	Common Frame of Reference
CID	Computer Integrated Documentation
CMap	Concept Map
CMMi	Capacity Maturity Model Integration
CONOPS	Concepts of Operations
CPDLC	Controller-Pilot Data Link Communications
CRT	Cathode Ray Tube
CSCW	Computer-Supported Cooperative Work
DAC	Dynamic Airspace Configuration
DARPA	Defense Advanced Research Projects Agency
DES	Discrete-Event Simulation
DGAC	French Civil Aviation Administration
DSC	Dynamic Stability Control System

EADS	European Aerospace Defense Systems
ESMD	NASA Exploration Systems Mission Directorate
ESP	Electronic Stability Program
EURISCO	European Institute of Cognitive Sciences and Engineering
FAA	Federal Aviation Administration
FDA	Food and Drug Administration
FIT	Florida Institute of Technology
FMC	Flight Management Computer
FMS	Flight Management System
FTP	File Transfer Protocol
GEM	Group Elicitation Method
GEOSS	Global Earth Observation System of Systems
GMO	Genetically Modified Organism
GPS	Global Positioning System
HCA	Human-Centered Automation
HCD	Human-Centered Design
HCI	Human-Computer Interaction
HFE	Human Factors and Ergonomics
HSI	Human-System Integration
HITLS	Human-In-The-Loop Simulation
HORSES	Human-ORS-Expert-System
ICAO	International Civil Aviation Organization
I&C	Instrumentation and Control
IFE	In Flight Entertainment
IHMC	Florida Institute for Human and Machine Cognition
IMC	Instrument Meteorological Conditions
ISO	International Standard Organization
JSC	Johnson Space Center
LCS	Life-Critical System
LER	Lunar Electric Rover
LKA	Line Keeping Assistant
MAS	Multi-Agent System
M&S	Modeling and Simulation
MIDAS	Man-Machine Integrated Design and Analysis System
MSP	Multi-Sector Planning
MBA	Master of Business Administration
NAS	National Aviation System
NASA	National Aeronautics and Space Administration
ND	Navigation Display
NPP	Nuclear Power Plant
NRC	United States Nuclear Regulatory Commission
NTSB	National Transportation Safety Board
ONERA	*Office National d'Etudes et Recherches Aerospatiales* (The French Aerospace Lab)

ONIA	*Office National Industriel de l'Azote* (French National Nitrogen Industrial Administration)
ORS	Orbital Refueling Systems
PAUSA	*Partage d'Autorité dans le Système Aéronautique* (Authority Sharing in the Airspace)
PRA	Probabilistic Risk Assessment
SBD	Scenario-Based Design
SCS	Socio-Cognitive Stability
SE	Systems Engineering
SESAR	Single European Sky Air Traffic Management Research
SEV	Space Exploration Vehicle
SIGCHI	Special Interest Group on Computer-Human Interaction (ACM)
SoS	System of Systems
SoSE	System of Systems Engineering
SPAR	Standardized Plant Analysis Risk
STCA	Short-Term Conflict Alert
STEAM	Science, Technology, Engineering, Arts and Mathematics
STEM	Science, Technology, Engineering and Mathematics
SWE	Software Engineering
TCAS	Traffic alert and Collision Avoidance System
TCS	Traction Control System
THERP	Technique for Human Error Rate Prediction
TMA	Traffic Management Advisor
TNT	Trinitrotoluene
TOP	Technology, Organization and People
TRC	Traction Control System
UML	Unified Modeling Language
URET	User Request Evaluation Tool
VHF	Very High Frequency
WYSIWYG	What You See Is What You Get

Chapter 1
Introduction

This book is the product of incremental syntheses of my courses in human-centered design (HCD) that I have prepared and given at the Florida Institute of Technology (FIT). These courses support the Ph.D. program in HCD that I also created in 2009 and am developing. Initially, my target was to create a program in aerospace human factors since I worked in this very specific domain for more than 30 years. For many reasons, I understood that the topic was broader than aerospace human factors and decided to focus on HCD in general. This book mirrors the program through my eyes. I used my background in human-computer interaction, cognitive science, mathematics and, of course, aerospace engineering.[1]

HCD is not a new approach to design, even if it can be nowadays very fashionable. Aerospace is a good example of a domain where participatory design[2] was fully integrated, involving experimental test pilots and aerospace engineers as well as many other actors. Times change and finance professionals now run many companies more than anybody else. These people do not have an interdisciplinary background, so they move design and engineering to lower priorities. It is time to better understand why this new way of doing things does not work, at least in the development of life-critical complex systems, and propose principles, methods and organizations that lead to human-centered sustainable products.

We need design thinkers who are natural leaders. We need leaders who understand **technology, organizations and people**. We need leaders who have the sense of innovation for the sake of humanity and not only for money. We need leaders who are able to overcome the current worldwide economy where syntax has become more important than semantics, and in the best case when semantics exists! We all need to understand what we want to do with technology; how we should organize ourselves towards a better life; finally finding out who we are and have become. HCD is being developed for all these reasons and issues.

---

[1] Some references will more specifically come from the series of International Conferences on Human-Computer Interaction in Aerospace (HCI-Aero) that I co-founded and co-chaired for 26 years (note that this series was called Human-Machine Interaction and Artificial Intelligence in Aerospace in the beginning).

[2] Participatory design strongly developed in Scandinavian countries during the sixties, and human-centered design as we know it today benefited from this school of thought and practice.

G. A. Boy, *Orchestrating Human-Centered Design*,
DOI 10.1007/978-1-4471-4339-0_1, © Springer-Verlag London 2013

Fig. 1.1 The TOP model

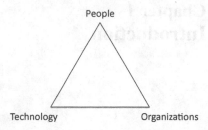

HCD has often been opposed to technology-centered design, or more precisely technology-centered engineering. Even if this distinction is relevant, I would like to introduce another distinction that is **finance-driven management**, to follow up what I just presented above. Engineers and designers have indirectly contributed to support the current dominance of finance-driven management. The spreadsheet, for example, was developed to automate calculations. Now, the spreadsheet and its extensions are widely and commonly used to manage small and large companies, to the point that people are reduced to spreadsheet boxes with a number in it! Everything becomes simple, when this number is a cost, it has to be reduced, and when it is a profit, it has to be increased... The initial purpose of the spreadsheet (i.e., calculation automation tool) has evolved to **organizational automation** leadership. We actually need to better understand what organizational automation means and why new types of human factors issues emerge from it. Workers are less motivated, involved, engaged, and increasingly incompetent. Finance-driven management drastically places money at the center of decision-making process and uses such an automated organization as an operational means; people and technology are most of the time at the periphery, where (re-)organization is used to incrementally fix issues, incidents and sometimes accidents. It is interesting to notice how many big companies constantly re-organize themselves to adapt to short-term requirements. This is mainly due to the fact that these companies are in an event-driven mode instead of a goal-driven one. Let us better understand these concepts.

The best way to improve our understanding and mastery of the evolution of technology, organizations and people (**the TOP model**: Fig. 1.1) is certainly to put to the front science, technology, engineering and mathematics (STEM) in a human-centered way (i.e., STEM should be intimately associated with history, philosophy, literature, and the arts). For that matter, Winston Churchill, in a 1948 speech in Oslo, Norway, said "Young people at universities study to achieve knowledge and not to learn a trade. We must all learn how to support ourselves, but we must also learn how to live. We need a lot of engineers in the modern world, but we do not want a world of modern engineers." Today, I would modify this statement by saying that "we need finance experts and accountants in our worldwide economy, but we do not want a world of bean counters, even the best ones." We need to involve humanities from the beginning: determining the need to evolve through focusing, and develop new philosophies with associated values based on our constantly evolving socio-technical world.

More than ever, I feel that our society has lost **educated common sense**. What does it mean? Educated common sense is simply doing things that make sense according to a well-understood purpose shared by an educated community of people. Back to my childhood in the countryside of the South of France when people were cooperating and sharing experience mostly by necessity, I remember how much these people developed educated common sense. This was not too long ago, perhaps 40–50 years! Today, when I go back to this region, the way of living is quite different; there are a few bigger farms that have become extremely mechanized. It appears that technology took the lead, and the population has dwindled. These people I have known are on the one hand more isolated, and on the other hand they seem to have lost this very precious educated common sense. When people need to work together, isolation factors require more coordination means that are not necessarily well developed today, even if the World Wide Web is defining new ways of coordination via social networks. This social interaction has become more mature. For example, we often tend to accept procedures and ways of life imposed by the current market economy without **critical thinking**. This is where educated common sense would be crucial.

The following is another good example. It is amazing to observe that a kiwi fruit from New Zealand that you buy on the Carmes market of Toulouse, France (i.e., approximately the antipode of New Zealand), is cheaper than a kiwi fruit that was grown in Montauban, a town located 50 km north of Toulouse. These fruits have to be flown from the other side of planet Earth, and specifically processed to resist to such travel. In addition, they are usually picked before they are ripe, which obviously decreases food quality in the end. Most people buy them because it is cheaper! Why can this kind of thing happen? This practice is not human-centered. It is finance-centered. It is interesting to observe that our global economy fosters such philosophy and practice. What could we do to reach a more sensible way of doing things in our socio-technical world? A solution is localizing instead of globalizing. More generally, when you want to make something great, you need to master its life cycle using human-centered principles and criteria. Only people working together in small teams, and more importantly know each other, can do this. It is not surprising that spreading-out processes among too many agents (e.g., companies) causes communication and production problems such as delays, mismatches, misunderstanding and interoperability issues. Making great things requires a continuum that our ISO[3] 9000 standards and quality processes do not have unfortunately. A good product requires care during its whole life cycle, and only well-coordinated competent people can do this; articulation work is at stake here.

What could be a solution for this kiwi fruit paradox? Apply the TOP model! As already said above, TOP means Technology, Organizations and People. First, technology should be designed and developed according to human-centered **purposes** (e.g., the purpose is eating good and healthy kiwi fruit). This is the technology part of the solution. Second, we need to move away from the finance-driven management model and bring human-centered design and management to the front. How can we overcome this contradiction between a global economy that is inherently complex, and the tremendously simplistic model of life that it attempts to support? I take the opposite view, which is harmonizing the complexity of life with many simple understandable economic models. Yes, our world has become more global

---

[3] International Standard Organization.

and complex because we are more interconnected than before at much larger scale. We need to understand differences between the beautiful natural complexity and the artificial complexity that we have built, too often with very short-term views. For example, farmers should be selling their fruits directly to customers instead of going through a long chain of resellers. This illustrates the current complicated organizational part of the solution. Next, we need to change our perception of risk taking and safety to install a safety culture that is based on a broadly educated background (e.g., people should be knowledgeable of what fruits are good to eat or not, and ultimately learn what is good for them and what is not). This is the people part of the solution.

Today, most managers are trained and act to take as little (financial) risk as possible. Of course, this is very valuable, but where is the beauty of creation and discovery in all of it? We have been (recently) trained to act in the short term and linearly. **Prediction is always short-term**; indeed we cannot predict what will happen in the long term. Prediction is always based on the notion of derivatives. In mathematics, a derivative is a projection in the very short-term. It enables us to simplify problems most of the time and build prediction models. Most importantly, these models enable us to react to events in real-time, when of course they are not too far from reality. For example, we have built autopilots on airplanes that have worked effectively and efficiently since the 1930s, but they do not work in the longer term, they correct small errors around a predefined steady-state trajectory. The pilot is therefore in charge of anything long-term! The short-term approach leads to **event-driven behavior** that is inherently reactive, tends to narrow focus and lose track of the overall goal. In contrast, leadership involves a **goal-driven behavior** (i.e., a good leader envisions **possible futures**, chooses one, and manages appropriate activities that will lead to realization of a human-centered possible future). In 1961, John Fitzgerald Kennedy made the decision to go to the Moon by the end of the sixties. NASA managed to implement this decision, with the appropriate means, and Neil Armstrong stepped on the Moon in July 1969. This success was the result of a real goal-driven HCD process.

It is time to come out from inside the box and investigate problems of the twenty-first century. The twentieth century was the era of **local and linear** problems. Engineering was centered on simplification of complex problems by decomposing problems into sub-problems that were easy to solve because they were modeled by linear approximations. These sub-problems were solved locally and solutions were re-integrated into the initial problem to be solved. This kind of approach worked very well most of the time because the number of components was small and interconnections were not so numerous. Today, our world has changed; the number of components is much bigger with more interconnections among them. We talk about systems of systems, and problems have become **global and non-linear**.

These problems are hard and complex, and they deserve more attention than they used to get until now. We then need to develop approaches, methods and tools to attack these problems. In this sense, HCD is becoming more than shaping a user interface on top of a system already built. HCD involves a broader perspective of what we want to do with technology, organizations and people (the TOP model again) to cope with current finance-driven organizational systems and to re-activate educated common sense in a project management leadership.

For that matter, this book is based on two main objectives: (1) investigate deeper knowledge in complexity science, cognitive science, organization science, life-critical systems, and other very important topics related to the evolution of our world; and (2) better understand educated common sense. The TOP model is at the center of these objectives, and promotes scenario-based design that requires both analysis of current events related to the use of technology, and anticipation of possible futures.

Complexity science is certainly becoming one of the most important sciences of the twenty-first century. We need to better understand why a system, made of simple entities very interconnected to the extreme, enables the emergence of very useful, useless or even catastrophic properties. On the one hand, we can be impressed by the way simple insects can organize themselves to produce extraordinary emergent behaviors. On the other hand, we need to understand the genesis of terrorism in order to eliminate or at least manage it. We need to better understand the way these multi-agent systems interact intrinsically and extrinsically, and figure out the emergence of properties of such systems that each agent did not initially have.

Life-critical systems (LCSs) can be artificial or natural systems. LCSs are systems where life is at stake. LCSs are always related to **risk taking**, which is a very important cognitive process in life in general. Risk-taking involves preparation, training and knowledge; it also involves situational awareness, decision-making and action taking. In the context of human-centered design, life-critical technology cannot be dissociated from the life-critical organization in which it will be used and the type of people who will effectively use it. The main trend in our occidental societies is protectiveness and the resulting notion of "zero risk". This is a change in mentality; the fear it engenders also tends to eliminate some life-critical activities that are crucial to human kind development. High-risk sectors such as aeronautics and space, and the nuclear sector, have developed **safety cultures** that have tremendously decreased the probability of fatal accidents. However, the risk of catastrophe still exists even if it is very small. Issues arise when there is a major accident. Communication and media take the lead and run the show. For example, between 1995 and 2000, there were 3 fatalities per 10 billion (3×10^{-10}) aircraft passenger miles, compared to 1.3 fatalities per 100 million (1.3×10^{-8}) car miles, which is 100 times higher.[4]

We really need to review our perception of risk, as well as the notion of risk itself. We need to review what we are generating for the next generations of people in terms of complexity and constraints due to the lack of awareness of what risk-taking really means. In particular, when we listen the current ambient philosophy of risk, people are always the problem; I strongly believe they can be the solution.

Talking about solutions, it is amazing to observe that, in our occidental societies, young people do not want to study **engineering** anymore. Engineering is not appealing because it is hard to study mathematics, physics and chemistry for example, and jobs do not pay very well in return for the investment. Many young people are more attracted by business because it is more straightforward and return on investment is

---

[4] http://en.wikipedia.org/wiki/Air_safety#cite_note-1.

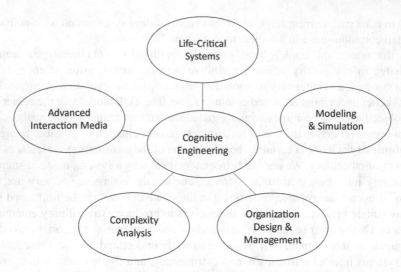

Fig. 1.2 Six themes supporting HCD of life-critical systems

very high. These are facts. However, we need to question what really needs to be done to save our current economical crisis. For that matter, organization design and management must be strongly revisited.

This is why new organizational models should be introduced, discussed and eventually implemented. This book first introduces the **Orchestra Model** (Chap. 2) from its initial genesis to explain the influence of advanced interaction media on our everyday life (Boy 1991), and later on within the context of the evolution of air traffic management in Europe, and generalized for life-critical systems (Boy 2009). It is interesting to observe that the traditional army model is currently replaced by the Orchestra Model (Boy 1991), where actors are more distributed, autonomous and specialized, and for that matter requires more coordination rules.

The Orchestra Model supports the entire HCD approach presented in this book, which is organized around six major themes (Fig. 1.2).

Chapter 3 gives an account on cognitive aspects of the way systems are designed, manufactured, certified and used (i.e., the reader is introduced to the growing discipline of **cognitive engineering**). The development of systems, where people are central, requires better understanding of safety, efficiency and comfort (i.e., we define **life-critical systems** through a series of domain examples such as aerospace, nuclear, automotive and medicine in Chap. 4). I already said in this introduction that complexity is key in this early twenty-first century; therefore **complexity analysis for human-centered design** is developed and illustrated (Chap. 5). HCD cannot be developed in an organization that is not appropriate, and consequently it is crucial to better understand **organization design and management** through currently developed models (Chap. 6). Since most products that we develop today are of high complexity, the maturity question deserves to be explored; for that matter, human-centered design cannot be done without **modeling and simulation** (Chap. 7), which provides appropriate means for continuous rationalization, evaluation and decision-making. In our information-based society, it is crucial to provide an articulated view on information technology and human-computer interaction, which I denote **advanced information media** (Chap. 8).

This following text is based on both a continuing reflection and real-world examples from a long practice in both industry and academia. I hope that it will motivate young people to become the designers of tomorrow and have an impact on current decision-makers and managers.

References

Boy, G. A. (1991). *Advanced interaction media as a component of everyday life for the coming generation. Proceedings of the World Marketing Congress.* Tokyo: Japan Management Association.

Boy, G. A. (2009). The Orchestra: A conceptual model for function allocation and scenario-based engineering in multi-agent safety-critical systems. *Proceedings of the European Conference on Cognitive Ergonomics.* Finland: Otaniemi, Helsinki area, (30 September–2 October).

Chapter 2
The Orchestra Model

Introduction

Technology is now almost always implemented with extensive software that enables humans to interact with machines, at least in very limited contexts. We commonly talk about **human and machine agents** (HMA). Human-computer interaction was traditionally thought as a person facing a computer in a one-to-one relation. Today, there are HMA networked societies as defined by Minsky (1985) (i.e., an agent being a society of agents). Human modeling, often commonly regarded in the information processing sense (Newell and Simon 1972; Wickens 1992), progressively migrated towards multi-agent organization modeling where cognition is distributed (Hutchins 1995).

In addition, the evolution of practices due to automation within an organization required developing appropriate models that can support human-centered design of multi-agent systems. The **cognitive function** concept (Boy 1998) emerged as a useful representation to support such modeling of both individuals (i.e., individual agents as organized structures of cognitive functions) and organizations (i.e., a set of cognitive functions distributed among a set of agents). Examples of vehicle control cognitive functions are "speed control", "trajectory management", "collision avoidance" and "failure detection and recovery". **Function allocation** in life-critical human and machine systems is not a new research topic (Fitts 1951). More than sixty years after Fitts's paper, there is a need for unifying what is separately done in design and engineering, human factors and organization science.

This chapter introduces the Orchestra Model that suits an evolution where agents require a common frame of reference (a music theory analog), contracts (scores) must be appropriately and formally coordinated (the role of the composer), real-time coordination must be assured (the role of the conductor), and an agent must have specific abilities to perform according to contracts (role and proficiency of the musicians). Contracts are seen as scenarios or storyboards, with an additional responsibility dimension. Consequently, the Orchestra Model will support the identification of both deliberate and emergent cognitive functions during the life cycle of a multi-agent life-critical system. It is based on previous work on cognitive function analysis (Boy 1998, 2002, 2007, 2011) and function allocation work (Grote et al. 2000).

G. A. Boy, *Orchestrating Human-Centered Design,*
DOI 10.1007/978-1-4471-4339-0_2, © Springer-Verlag London 2013

This metaphoric model was motivated by the introduction of new concepts of operations (CONOPS) in air traffic management (ATM), such as task delegation from ground controllers to flight crews. This transfer of **authority** is expected to induce emergence of new cognitive functions among the various ATM agents whether humans or automation. More specifically, the current hierarchical model of air traffic control (ATC), where authority is centralized on the ground, is evolving toward a distributed model of authorities that need to be coordinated among agents, whose functions are also evolving. Authority has become the central driving force of the Orchestra Model, where authority encapsulates both control and accountability.

From Technological Automation to Organizational Automation

During the twentieth century, we substantially improved material comfort both at work and at home, and more generally in our everyday life (e.g., transportation is faster and safer, power supply is available almost everywhere, our kitchens are well equipped, we have efficient washing machines, entertainment systems and so on). More recently, the Internet was massively integrated into our lives. It considerably improved communication among people, as well as information access. Everything is at our fingertips. But, is our life better? Obviously, insiders and technology fans manage to use the Internet appropriately for example. I personally find the Internet very liberating and extending our personal knowledge. I think the Internet has the potential to make us smarter. However, for many people technology may not make their life better. Why not? Most of us do not take or have time to wait for suitable maturity of both technology and related practices. Technology and practices are constantly co-adapted, and consequently this co-adaptation extends to the organization. Let us start to review what has been done in technological automation before focusing on organizational automation.

Technological Automation and the Concept of Cognitive Function

The clock is probably the oldest artifact that humans have automated. Today, nobody questions its use. We wear watches to know time like we wear clothes to keep warm. The first autopilots were installed in commercial airplanes in the 1930s. Pilots learned how to use them. Automation theories were incrementally developed from various disjoint domain backgrounds, including James Watt's regulator for the automatic control of a steam engine at the end of the eighteenth century, and Claude Bernard's theory of the controlled stability of the *internal milieu* (Grande and Visscher 1967) at the end of the nineteenth century. It took the contributions of biologists, engineers, mathematicians, physicists, and more recently computer scientists (not an exhaustive list) to develop the automation that we have today. But in most cases, human factors were not sufficiently anticipated during automation design. The use of automation was not anticipated enough, and the emergence of new practices not expected.

Fig. 2.1 A cognitive function as a transformation of a task into an activity

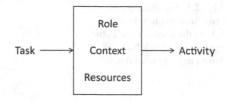

At this point, the *cognitive function* concept (Boy 1998) needs to be explicitly defined. A cognitive function is a representation that was developed to support the analysis of human-computer interaction in aeronautics. A cognitive function is defined as a cognitive process that enables the transformation of a task (i.e., what is prescribed and should be done) into an activity (i.e., what is effective and really done). It is typically described by three properties: its role, its context of validity and a set of resources (Fig. 2.1).

It is a practical representation since it can support the description of both human and computer (machine) functions. In Fig. 2.2, circles represent human cognitive functions and squares represent machine cognitive functions. In particular, automation can be defined as cognitive function transfer from human to machine (Fig. 2.2).

Note that both the human and the machine can be represented as "high-level" cognitive functions, which are themselves organizations of cognitive functions.

I define **human-centered automation** as the transfer of one or several cognitive functions from a human to a machine, augmented by the cognitive functions that emerge from the *use* of the resulting automated machine (Fig. 2.3). As already stated these cognitive functions may emerge as surprises.

Many **emerging cognitive functions** cannot be anticipated at design time, they can only be discovered at use time. This is why we cannot fully eliminate surprises in practice. Of course, human-in-the-loop simulations facilitate such discoveries and should be systematically carried out before a new (automated) product is delivered. Afterwards, systematic experience feedback completes the quest for emerging cognitive functions. Obviously, these must be shared within the community of practice to facilitate their validation and to empower the overall community.

For the last three decades, we have investigated the shift from technology-centered to human-centered automation design (Billings 1991; Boy 1998). It is crucial to anticipate the consequences of wrong practices in safety-critical human-machine systems.

Fig. 2.2 Automation as a human-to-machine cognitive function transfer

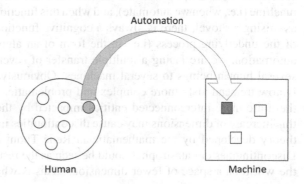

Fig. 2.3 Automation as
cognitive function transfer,
where the central issue is the
systematic search for
emerging cognitive functions

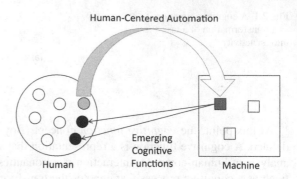

First, we need to know as many of these consequences as possible. However, we also need to be aware that we will not be able to know all of them. When consequences of this incompleteness are not important, automation can make itself invisible to its users, but when consequences may become catastrophic, users must know what the automation is doing and measure risks when they are using an automated system. They must know how it works, both externally and internally, as much as possible. This is where risk-taking enters into play. This is made more complicated because automation is usually not just a matter of a single human operator interacting with a sophisticated machine. It is often a matter of several human and machine agents interacting within a complex socio-technical organization. Automation problems arise when there is not enough maturity of practice within a real organization (i.e., involved users are not aware of what they do using automation in their everyday work or life). In life-critical systems, to be effective, people involved must be knowledgeable of what the automation role is, as well as its context of validity and the resources required. The big question that remains is the level of granularity of this knowledge (i.e., what degree of details users should know about the automation that they are using). Before discussing socio-technical organizational complexity in more depth, let us develop the concepts of organizational automation and authority.

Organizational Automation and the Concept of Authority

Whenever we attempt to transfer one function from a "generic" human being to a machine (i.e., when we automate), and when this function is a physical function, such as raising a shovel, there is always a cognitive function that enables the execution of the underlying process (i.e., in the form of an algorithm). With organizational automation, we are facing a multiple transfer of several cognitive functions from several human beings to several machines. Obviously, the concept of automation is now tremendously more complex and problematic. It is more complex because there are more interconnected entities impacted by the automation, and therefore this increase of dimensions may cause discontinuities in the sense of the catastrophe theory developed by the mathematician René Thom in the 1960s (Thom 1989). Discontinuities or catastrophes could be seen as the result of a specific projection of the world on a space of fewer dimensions. This is what users or human operators

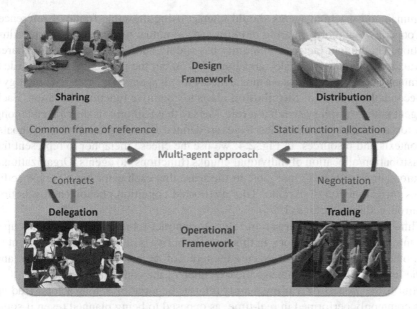

Fig. 2.4 Authority sharing, distribution, delegation and trading. (Stock images courtesy of liquidlibrary/Thinkstock, Photos.com/Thinkstock, AbleStock.com/Thinkstock, Hemera/Thinkstock)

do when they try to simplify the complexity of the automated system they have to manage onto a smaller dimension complexity space in order to maintain reasonable situation awareness. It happens that this kind of projection creates accidents.

Computers and software enable tremendous possibilities in information processing and management. Today, almost all work systems are software-based. With these new possibilities, machines emerge as real effective actors in the overall human-machine systems (i.e., software provides them with special kinds of cognitive functions). In theory, systems should provide more control and planning capacities as well as greater flexibility.

In practice, this multi-automation evolution brings to the forefront issues of **authority** because there are more situations where we do not know who is in charge anymore. Authority can be defined as both control and accountability (Boy and Grote 2009). We started to study these notions in the national PAUSA project (Boy et al. 2008). PAUSA literally means *"Partage d'AUtorité dans le Système Aéronautique"*, a French acronym for "Authority Sharing in the Airspace". Nine major organizations, twenty-six researchers and practitioners were involved. We deliberately focused on the identification of human and organizational factors involved in ATM automation in 2007, in 2015 and after 2020. One of the major results from this project was the definition of four main processes that support authority allocation in organizational automation (Boy 2009). These processes are presented in Fig. 2.4 and explained as follows.

- First, the process of **sharing** refers to information, knowledge and resources involved among a set of actors who need to have a shared awareness of the situation.

Human and machine agents should interact using the same frame of reference. Cooperative authority sharing requires such a mutual understanding of the situation. The human factors community has widely addressed this issue of shared situation awareness (Endsley and Jones 1997), but the point here is to provide a framework of reference, like a music theory for all agents (i.e., a shared ontology).

- Second, the process of **distribution** refers to cognitive function allocation. Each agent in the overall system has a role. We say that authority is distributed among actors with respect to their roles. Roles are defined according to competency, goals, contexts and resources. In Fig. 2.4, we use the cheese metaphor to represent the distribution/allocation of individual chunks (functions) to agents. Organizational automation tends to format jobs, in the sense that each actor has a very specific cognitive function to execute. This distributed cognition phenomenon was first introduced by Hutchins (1995).

- Third, the process of **delegation** refers to contract setup (i.e., an actor sets up a contract with another actor). In this case, authority is delegated from an agent to another (i.e., the latter will execute the terms of the contract for the former, and the former will have to manage the execution of the contract).

- Fourth, the process of **trading** mainly refers to negotiation among actors. Trading is commonly performed in real-time, as opposed to being planned (even if some patterns could be anticipated in order to facilitate trading operations), and requires the best mutual understanding among competent actors. Authority is traded among actors according to possibly conflicting goals (or roles), contexts and resources.

Organizational automation is designed incrementally from the early stages of the design and development processes to actual operations. Constructing frame of reference (the sharing process support) and static function allocation (the distribution process support) could be started very early. Delegation and trading processes require operational setups because unanticipated cognitive functions will emerge from practice. This is why human-in-the-loop simulations are necessary during organizational automation's development phase. Problems arise from task definition, coordination and ability of all actors to articulate their work with others. Even if each actor has a well-defined cognitive function, its context of use is often too rigidly defined for appropriate articulation and intersubjectivity[1] among actors. This is why a model that enables the study of socio-technical evolution and emergence of new practices is critically needed.

There is a vast amount of research in the domain of shared (or not shared) mental models (Kanki and Foushee 1989; Walz et al. 1993; Malone and Crowston 1994; Faraj and Sproull 2000; Mathieu et al. 2000). However, there is still a large amount of research to be conducted on the difficult problem of representing and simulating interactions among agents. Awareness of the organizational environment is crucial here. Two human beings interacting between each other incrementally **adapt** their own awareness of the other as conversation goes on (i.e., each of them learns from

---

[1] Intersubjectivity typically refers to a psychological relation between people, whether it is common sense, an agreement or a divergence.

the other and consequently his or her model he or she has of the other evolves during conversation). Interaction with or through machines is a different matter. We will develop this question in the form of three models of interaction later in this chapter.

Function Allocation

Paul M. Fitts edited a famous report on human engineering in 1951, where he and his colleagues drafted possible roles of the human operator in future air-traffic control and navigation systems. Fitts and his colleagues developed principles and criteria to design and assess the division of **responsibility** between human operators and machines, as well as among human operators themselves. They anticipated issues in decision-making, the nature of information, the form that information may take (i.e., encoding), the rate of flow of information, its storage, perturbation, redundancy, and related research problems. They mostly focused on visual and voice communication problems. Among other things, this report provided what is now known as the Fitts's list of where humans appear to surpass machines and conversely (Fitts 1951). This preliminary work led to several lists of strengths and weaknesses of human operators and automated machines (Chapanis 1965; Swain and Guttman 1980; Sheridan 1987). They were called MABAMABA (i.e., "Men Are Better At—Machines Are Better At"). This was an easy but very limited way to provide guidelines for automation design (Parasuraman et al. 2000). Later on, Hollnagel and Woods (2005) based their approach on the fact that joint cognitive systems (humans and machines) are dynamic and therefore complex, and need to cope with this kind of complexity at both individual and organizational levels. This approach is descriptive and requires operational developments.

Function allocation cannot be only addressed from a static point of view; it can also be highly dynamic, because underlying processes are dynamic. In this case, it would be better to talk about real-time function adaptation, even if this is often referred to as dynamic function allocation (Corso and Maloney 1996; Hildebrandt and Harrison 2003). Function allocation can also be dynamic because cognition is distributed among a set of agents (Hutchins 1995; Wright et al. 2000).

Of interest is the fact that the next generation of ATM systems will have to be designed taking into account principles and criteria for both static and dynamic function allocation. What drastically changes today is magnitude of the air capacity (i.e., the number of aircraft is tremendously more than in 1951). Consequently, a conceptual model that is able to support the study of the next generation of ATM systems cannot be based on a single agent approach, but needs to be based on a multi-agent approach from the start. It is no longer possible to analyze each agent in a system individually because interrelations are far more numerous and complex than before. Technology is information intensive and organizational setups need to be revisited. Furthermore, agents are no longer only human operators, but also automation in the form of various new kinds of software agents dedicated to specific tasks. For that matter, function allocation cannot be thought as an a priori process, but as an evolutionary process. Function allocation among a set of numerous interconnected agents is a difficult problem, because, like a multi-agent biological entity, complexity is as much in the links between agents as in agents themselves. When number of agents and interconnections among these agents increase, function allocation will increasingly become dynamic, managed and optimized by specialized human or machine agents. The evolving ATM system is an excellent example of such multi-agent systems. As in biological systems, **separability** is an issue (i.e., it is difficult, and often impossible, to separate a part of an overall system and study it independently of the rest).

It is crucial to maintain awareness of internal constituency of life-critical human-machine systems as well as external threats. Within such multi-agent systems, it is crucial to know which agent does what and when. Like in a flock of birds, each bird has basic functions that enable it to ensure the stability of the whole flock. In addition, each agent should have the capacity to execute the task he/she/it has to perform in the right context (i.e., an appropriate function should be allocated to this agent). Each function allocation has a cost that should be carefully understood and eventually measured. Finally, each function allocation induces a level of **confidence** and **trust** in the agent (Campbell et al. 1997). When the agent is a human, function allocation can be defined in terms of level of training and experience. When the agent is an automated system, trust can be characterized by several metrics such as reliability, flexibility and cognitive stability. Cognitive stability was defined as the ability to recover from human errors or system failure (Boy 2007). We will further discuss cognitive stability in Chap. 5.

A cognitive function could be defined recursively by a network of cognitive functions that may be distributed among various agents in an organization, and across various meaningful contexts of use, whether nominal or off-nominal. Any time a new cognitive function is defined or moved from one agent to another, it is crucial to look for new cognitive functions that emerge from various interactions in the related agent network. Human-centered automation is an incremental process where design, test, practice and discovery of emergent cognitive functions are intertwined. This approach to automation is strongly based on operational expertise, development of scenarios, human-in-the-loop simulation (HITLS) and formative evaluation. For a very long time, the aerospace industry used this type of approach where flight tests were an essential part of the engineering process.

Scenario-based Design

Scenario-based design (SBD) changes the focus of design work from defining system operations (i.e., functional specifications), to describing how people will use a system to accomplish work tasks and other activities (Carroll 1995, 2011). SBD is necessary prior to systems engineering activities. If systems engineering provides syntax, SBD provides semantics of the problem to be solved. SBD elaborates a

traditional principle in human factors and ergonomics (i.e., human attributes and requirements should be taken into account in design and development). SBD consists
of describing usage situations as design objects. It starts with involvement of appropriate domain experts. In the PAUSA project, pilots and ATC personnel were the
domain experts. During early phases of design, envisioned scenarios were developed
from expert knowledge and knowhow. Such scenarios are usually built as extensions
of current observed scenarios in the real world. They may be designed as analogs of
similar **configurations and chronologies** observed in other domains. Scenarios are
constantly readapted to support human-centered design appropriately.

In the context of function allocation for life-critical multi-agent systems, such as
construction of commercial airplanes, the need for scenarios takes on another dimension. Even if it is clear that scenarios are needed from the beginning of the design
process, they are also needed along the entire product's life cycle. Scenarios not only
serve as storyboard guides in human-centered design and development; they also
provide an excellent framework for rationalization of evaluation-test results, which
in turn are used for re-engineering the product and improve its safety, usefulness
and usability. For that matter, SBD supports function allocation during the entire life
cycle of a socio-technical system.

A distinction is made between declarative and procedural scenarios (Boy et al.
2008; Straussberger et al. 2008). Such distinction is not new since it was used for a
long time in artificial intelligence, and more precisely in logic programming (Mc-
Carthy 1958). In cognitive psychology, this distinction usually refers to declarative
memory that involves conscious recall of facts and events, versus procedural memory that is about unconscious acts, reactions or skills (Ullman 2004). In software
engineering and system engineering, another type of approach was developed called
"use cases" that enables representation of functional scenarios (Cockburn 2001). The
Unified Modeling Language (UML) was developed for this purpose also (Jacobson
et al. 1998).

In this chapter, scenarios are thought of in the same way as movie scenarios.
Declarative scenarios describe necessary objects and agents involved in the final
product. These objects and agents are presented in the form of structure and function. Such descriptions lead to the way objects and agents interact with each other,
and subsequently to application use cases. **Procedural scenarios** describe event
chronologies and interactions among objects and agents. Such descriptions are stories and episodes that lead to appropriate definitions of such objects and agents.
Declarative and procedural scenarios may be initially developed by different groups
of people in isolation. These two types of scenarios are developed concurrently to
improve completeness of both objects/agents and their possible interactions. They
are incrementally merged into synthesized generic scenarios.

Technical systems should never be looked at in isolation, but always as part of
a bigger socio-technical system, which includes humans operating the system as
well as formal and informal structures and processes within which they work. This
is why scenarios are so important because they support rationalization of meaningful interactions in the socio-technical system being studied. There are situations
that are very difficult and even impossible to predict before they actually happen.
These situations are usually called surprises. For example, the 2002 mid-air collision

accident at Überlingen Germany, which has been extensively analyzed (Weyer 2006), has shown that introduction of the Traffic alert and Collision Avoidance Systems (TCAS) as a deconstruction of order or even a regime change, may be a gradual shift from central control to decentralized self-organization. Some accidents such as the Überlingen one highlight such evolution and sometimes revolution, in the overall socio-technical system where coordination has become a major issue. This is why we deliberately choose a multi-agent approach, instead of a single-agent approach (e.g., the pilot facing a screen), to express function allocation. To do this, we need to develop a common frame of reference, task delegation, and information flows among agents. ATM evolution is seen as a shift from army to orchestra from recent experience-based investigations (Boy and Grote 2009; Boy 2009). This metaphor and its possible extensions support the airspace multi-agent system evolution. In this way, we expect to move from the clumsy design-and-surprise approach to a principled design approach based on SBD toward the development of more robust and resilient socio-technical systems. We argue that such a principled approach to design could have avoided the Überlingen accident, by recognizing that a connection between TCAS and ground-based STCA (Short-Term Conflict Alert) systems could have afforded much better controller's situation awareness.

During the last decades, human factors researchers blamed engineering for waiting for surprises in order to correct ergonomics of products (i.e., structures and functions of products). Unfortunately, there will always be **surprises**. The main issue is to try to anticipate them as much as possible before the final product is delivered. It is difficult to imagine other ways than constantly developing deeper knowledge and knowhow from positive and negative experience by using the product. Our technological society is developing very fast, tremendously faster than before. We do not take enough time to analyze our mistakes and generate syntheses (e.g., in the form of "golden rules"). Scenarios are good tools to pose questions such as Who, What, When, Where, Why, How and How much (5W2H): Who and What are the agents and objects and relationships among them along relevant dimensions such as time (chronology), functional and structural dependencies, topological organizations and so on. Why do these agents exist in terms of role, context and resources (i.e., cognitive functions)? When and where are they useful, active or potentially related to one another? How do they work or how can people work with them or use them? How much is load involved in user's activity, in terms of workload, appropriate cost of any kind and so on? This is why maturity has become a field of research that is far from being mastered and "mature".

SBD should then look for maturity and flexibility. The issue of **maturity** has been analyzed before (Boy 2005). We know that we must focus on product maturity and practice maturity (i.e., what the product is for and how it is really used). **Flexibility** is also a key issue (i.e., products must be adaptable). For example, in the ATM with a constant increase of aircraft in the sky, planning is required but must be considered as dynamic planning that provides flexibility in case of unexpected events.

Product maturity and flexibility are strongly based on quality of **high-level requirements** and their constant adjustments to the real world during the whole life cycle of a product. Of course, if high-level requirements are not right or strong enough in the first place, chances are that designers and engineers will have to re-engineer the product many times in the future, and sometimes get rid of the product unfortunately. This is why **starting right** is the best advice we could give to a design team. But what does it mean to start right? It means starting with appropriate scenarios, and consequently the best high-level requirements. In addition, it means the product should be regarded as a global entity and not a juxtaposition of pieces that will eventually be assembled. This is another reason why human-in-the-loop simulations are crucial as early as possible during the design/development process. Finally, teamwork must be cohesive from the beginning of design to operations and until obsolescence of the product. Following up on this analysis, there is a need for a conceptual model that could support function allocation, SBD and reaching maturity. This model is presented in the next section.

The Orchestra Model

The Orchestra Model requires the definition of the authority concept. It was designed over the years (Boy 1991) and finally refined during a study carried out from 2006 to 2008 on **authority sharing** in the airspace system, the already mentioned PAUSA project. Authority is defined from two main perspectives:

- **control** in the engineering sense (i.e., who is in charge and competent for a given task and function), and
- **accountability** in the legal sense (i.e., we are always accountable to someone else, and accountability includes responsibility).

Results were based on the definition of a scenario-based approach that supports the design of such HMA systems. We used declarative and procedural scenarios. The main problem was to obtain meaningful and generic scenarios that would be the source of emergent cognitive functions during further human-in-the-loop simulations (HITLSs), which range from very simple paper and pencil narrative storyboards to the use of interconnected very sophisticated realistic simulators. Both involve domain experts.

In the beginning of a project, as during its whole life cycle, strong expertise and experience from operational practitioners is required to develop useful scenarios. ATM practitioners are pilots and ATC controllers (ATCOs), as well as aerospace designers and certifiers. It is of course also important to motivate and carefully filter their inputs in a reflexive way through creative designs. However, testing will always remain the mandatory (long and recurring) process in which design and development processes will have to comply.

Music Theory, Composers, Scores, Conductors, Musicians and Audience

No simulation can be purposefully and efficiently carried out without a **conceptual model**. In this SBD approach to function allocation, the Orchestra Model is an alternative to the traditional army-type model that supports a hierarchical decomposition of functions. Five categories of entities must be defined.

- First, the **music theory** that supports various information-flows and provides a common frame of reference for all agents in the environment.
- Second, the **scores** that agents are required to use in order to support their assigned functions during operations. **Composers** typically develop scores and articulate them among each other to coordinate overall performance of the musicians. These composers still remain to be identified correctly in the ATM case.
- Third, **conductors** who provide the operational timing patterns, and consequently will be responsible for effective information flows (i.e., the overall symphony performance) among musicians and the audience (i.e., end-users).
- Fourth, **musicians** themselves who are required not only to perform what their scores say, but also to articulate their own performance with the others.
- Fifth, the **audience** that includes customers of the symphony.

Supervision, Mediation and Cooperation by Mutual Understanding

In an HMA organization such as an orchestra, agents are interrelated with respect to three kinds of interaction models (Boy 2002). These models are distinguished by knowledge each agent has of others in the organization.

1. When agents do not know each other, the best way to interact safely, efficiently and comfortably is to be supervised. **Supervision** is the first interaction model. None of the supervised agents has the authority to decide what to do; a supervisor does it for them.
2. **Mediation** is the second interaction model. Agents have a common frame of reference (CFR) through which they are able to interact. They still do not know each other deeply, but they know that they can interact between each other through the CFR. In addition to CFR, there are mediating agents who facilitate interactions. In WYSIWYG user interfaces for example, in addition to desktop metaphors, there are mouse-sensitive help lines that pop-up on demand. In this model, authority is distributed among the agents.
3. The third interaction model is **cooperation by mutual understanding**. This is what people usually do when they interact with each other. This model assumes that agents are able to construct a mental model of the others in order to perform better in future interactions. People interacting among each other do this naturally. Very simple instances of such a model have been developed and used so far on computers. For example, some pieces of software are able to learn user's habits and incrementally provide smart options or suggestions. This is the case of current text processors that are able to learn a user's specific lexicon from frequent uses of words. Web browsers remember frequently used links, etc. In this model, authority is traded between the agents. In human-human interaction via machine agents, related technology should provide

Fig. 2.5 Interaction models from no-autonomy to full-autonomy of agents

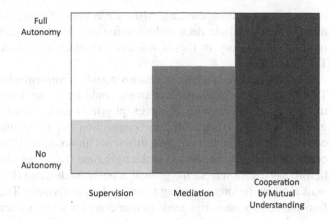

appropriate situation awareness means to enable sustainable and symbiotic communication.

To summarize, there is a continuum from the supervision model of interaction where authority follows a top-down army-type model, to the mediation model of interaction where authority follows a transversal orchestra-type model, to cooperation by a mutual understanding model of interaction where authority follows a more-chaotic trade model (Fig. 2.5). These interaction models are very useful to support the way cognitive functions are implemented in complex software not only from a human-computer interaction point of view, but also from an internal subsystem-to-subsystem point of view. In particular, they also provide an articulated way to validate large object-oriented software.

These three interaction models implicitly assume that agents have the same global goal or local goals compatible the global goal (e.g., performing the symphony). It may happen that some agents have competing goals (e.g., competing for a limited set of resources). Competition is an orthogonal dimension to the three interaction models where supervisor would become referee, mediator would become counselor, and competing agents could mutually understand each other to improve gains. Therefore, the three interaction models can be non-competitive or competitive.

Function Allocation, Scenario-based Design and Maturity Reaching

How can the Orchestra Model support function allocation, scenario-based design and maturity reaching? First, functions are allocated with respect to agents' competence and availability, as well as their interdependence defined by the overall work organization. For example, the collision avoidance function can be allocated to pilots, if their aircraft is TCAS-equipped and they know how to use TCAS (availability and competence). Also, some kind of coordination should be implemented between flight deck and air traffic control (ATC), either managed by people (resolution advisory in the TCAS case) or machines (direct connection between TCAS and ATC).

Second, several types of scenarios could be implemented and assessed. In the TCAS case, two declarative scenarios could be set up defining two major configurations: (DS-1) one with no direct (physical) connection between flight deck and ground; and (DS-2) one with direct connection. In parallel, two procedural scenarios could be developed defining two different chronologies: (PS-1) one in which pilots communicate with the ground at the right time (i.e., resolution advisory is provided before any decision from the ground is being made); and (PS-2) one in which pilots wait too long before providing their resolution advisory. The combination of these four elementary scenarios leads to four complex scenarios that provide a good basis for evaluation of function allocation. In this case, the combination of DS-1 and PS-2 is likely to bring surprises such as the Überlingen accident.

Third, the multi-agent system becomes mature when each agent understands its role in the right context using available resources (i.e., cognitive function definition). Maturity is guided by the type of interaction agents that are able to perform in a given context (i.e., supervision, mediation or cooperation by mutual understanding). In particular, it is clear that whenever cooperation by mutual understanding is possible, interactions are very advanced among agents, but we also need to know when human and/or machine agents need mediators, and even more specifically supervisors. Maturity is then tested in terms of level of autonomy (i.e., an agent is said to be autonomous when he/she/it is able to handle its allocated function independently). In some cases, an agent is not able to be autonomous and requires assistance in the form of mediation or supervision. It is important to realize soon enough during the design process when this kind of interaction model might be predominant. Surprises arise when this kind of test is not performed in advance. Note that supervision and mediation are perfectly acceptable models when they are well understood and accepted by the various team players. Like musicians who are collaborating within an orchestra, agents need to accept either the lead from other musicians or to be mediated through scores, which themselves were coordinated by the composer. The maturity process may take some time before the multi-agent system reaches an acceptable maturity level. It is based on competence of each agent, their abilities to work with others in the "orchestra" and a clear understanding of the various interaction models that are required to run the overall "symphony".

Authority Sharing Illustrated in the Aeronautical Domain

Authority sharing is one of the major themes of the next generation of air traffic management system, and flight deck automation in particular. The fact that we will have more aircraft in the sky (i.e., air traffic capacity increase), and we want to enhance safety, requires deepest research on the way various functions are being reallocated among various agents. We need to better identify pilots' information requirements and communication needs to perform tasks currently managed by air traffic control, which will greatly increase the needs for a pilot's awareness of the surrounding airspace, (human and system) failure identification and recovery, and unexpected-event handling in this dynamic and complex multi-agent infrastructure.

Therefore, we need to co-design and co-adapt both technological and organizational support. Avionics software is now highly sophisticated, enabling many machines to be considered as agents (i.e., having cognitive functions as humans have). Human and machine agents are more interconnected in the airspace than ever before, and their inter-relationships are often crucial to understand, master and support. This evolving ATM multi-agent world is critically situated and context identification is a primary concern. In particular, flight deck automation will have to be designed taking into account that pilots will gain autonomy, thus changing coordination requirements.

Consequently, function allocation needs to be addressed during the whole life cycle of all ATM systems. Cognitive function analysis is typically used to support the analysis, design and evaluation of such function allocation. More specifically, cognitive processes, such as authority sharing, distribution, delegation and trading, must be addressed. While there are human cognitive functions that can be predicted during design, there are some that will only emerge from use. This is why scenarios should be extensively developed and HITLS carried out.

Merging and spacing in dense traffic is a difficult problem; in particular, the sequencing of arrival flows through a new allocation of spacing tasks between air and ground. Today, ATCOs solely manage aircraft merging and spacing in busy airspaces. They control both the sequencing decisions and manage the merging routes, airspeeds and altitudes, guiding each aircraft. Controllers are aided by today's tools, which range from simple Letters of Agreement (LOA) and standard navigation aids, to more advanced systems like today's GPS approaches and integrated Flight Management Systems (FMS). The new Required Navigation Performance (RNP) procedures are the latest improvement down the traditional path of providing pilots with standard procedures and more accurate ways to follow them. While this approach is an important one, it alone will not solve the future problems of airspace congestion because it addresses only execution and does not address the major issue, which is coordination. Today, ATC is a centralized army-type decision point (i.e., all decisions must pass through this point and be distributed in a serial manner to all pilots within the managed airspace). This is a clear bottleneck that is highly dependent on skills of the controller to analyze the situation, make decisions, and then communicate required information to each aircraft as necessary.

Pilots under instrument meteorological conditions (IMC) have traditionally been "flying blind" with respect to other aircraft around them. Good pilots typically build mental maps by listening to the radios (party-line) and piecing the scene together. Recently, TCAS and Automatic Dependent Surveillance-Broadcast (ADS-B) have started providing pilots with a little more awareness of their immediate environment. These technologies provide the pilot with a "second set of eyes" besides the controllers. This information allows pilots to make decisions of their own, but unfortunately it is not coordinated with ATC, which has resulted in unfortunate accidents, again highlighting the importance of coordination.

Future ATM systems will enable pilots to be more autonomous and consequently will require more coordination among agents. They will have contracts like musicians have scores. Consequently, these contracts will have to be coordinated by some sort of planners, like the composers do. From this point of view, the main difference between ATM and a symphony is that contracts may change during performance, like the play of a Jazz orchestra. Authority trading will be a major issue. Situation awareness of each agent remains a central emergent cognitive function to investigate and identify during design and development. In fact, agent's authority and situation awareness are intimately coupled, and their identification determines the type of interaction model an agent will have with other agents that are relevant in the operational context. Sometimes supervision is the only interaction model that is possible, and agents will need to refer to a conductor. In other situations, they will be able to interact via contracts (scores) and trust this mediating means. Finally, it will happen that agents will perfectly understand what the others are doing, and therefore will communicate directly.

Off-nominal situations are infrequent, but have a tremendous impact when they do occur. They typically induce dynamic function allocation:

- appropriate agents will have to be aware of the situation change (resulting in a different common frame of reference, or music theory and style if we take the Orchestra metaphor),
- contracts will have to be redefined and coordinated (composer role), and
- consequent operations will have to coordinated (conductor role).

For example, it may happen that an aircrew is not able to make it off the runway at the high-speed exit and take a full-length landing. In a congested terminal area, the following aircraft will have to perform a go-around maneuver. First, the aircrew must realize they are not going to make the exit (situation awareness cognitive function), they must manage the landing (safety-assurance and action-taking cognitive functions), and find time to let the controller know (coordination cognitive function). Consequently, the ATCO must inform the trailing aircraft and potentially all other aircraft sequenced on the approach (coordination cognitive function). All these cognitive functions must be implemented at the right time, which might not be the case given the extra workload during this kind of operations. Information flows are highly dynamic and can only be managed by well aware and knowledgeable agents, and possibly new technology. For example, the ATCO re-sequencing traffic may also discover there is an aircraft that is low on fuel and requires an emergency landing. Creative decision-making is consequently the appropriate cognitive function that is at stake for the ATCO. On this very simple example, we see that authority must be shared in a timely manner among appropriate agents.

One way of managing this coordination problem is to develop appropriate automation. Automation can be used to detect when an aircraft will not make the exit and automatically signal the controller, elevating this burden from the pilot who is likely under high workload already. That same signal could automatically be sent to all the trailing aircraft. This kind of additional agent is expected to create more situation awareness among involved agents and therefore increase their mutual understanding of the situation (thus promoting the third interaction model). In addition, the ATCO, as a conductor, could make a single call confirming the situation and requesting reduced speeds. Each aircraft could acknowledge through their flight displays instead of using radio communications and ATCOs would see each response on their own screens. If this kind of solution seems to simplify the job of the various agents, it is mandatory to make sure that they are properly trained or fine-tuned, using the right cognitive functions.

In these examples, we can see that cognitive function analysis using the Orchestra Model enables investigation of the various relationships among agents and emergence of new cognitive functions, such as appropriate automation. Of course, any solution needs to be tested and further validated in HITLS or in the real world.

Systems of Systems

A system of systems (SoS) is an integrated set of many self-contained independent, task-driven or domain-specific systems that cooperate by combining their roles and resources. They work in specific contexts, to satisfy a global need and enable the emergence of a new, more complex, super-system that offers greater performance and emerging functionality than simply the sum of the constituent systems.

For example, our evolving air traffic management system is a system of systems where various agents constantly redefine their roles, interconnections and interdependencies. Authority sharing has become a crucial issue because both ATM automation and organization constantly evolve. One main reason is the constant increase in number of aircraft and the necessity of improving safety.

The concept of SoS is now widespread in the engineering of contemporary systems.[2] The SoS research discipline is currently evolving by developing frames of reference, thought processes, quantitative analysis, tools, and design methods.

How does a SoS differ from a mechanical system made of many components (e.g., an old clock)? SoSs denote large-scale complex systems where properties are **emergent** and can only be identified at the time of use. It is therefore important to discover these properties and their emergence as early as possible. Indeed, building a SoS is a holistic endeavor that requires architects and appropriate tests to foster emergence of salient properties and hidden cognitive functions. More generally, the concept of super-system emerges from interaction among its components.

Designing, engineering and manufacturing a commercial airplane is today a SoS endeavor. The way the company is designed and orchestrated is crucial. At the composer level, segmentation of work should be carefully thought not only in terms of processes to be implemented but also in terms of articulation among these processes. As in a symphony, timing is a major issue in terms of synchronization, mutual understanding and integration of a successful product. No matter how a company is initially structured with respect to safety, efficiency and comfort (this is the role of

---

[2] The following is based on our experience in the design of air traffic management systems. The author was the coordinator of the PAUSA project (already mentioned). PAUSA emphasized the concept of systems of systems in the aeronautical system, and more specifically the issue of authority among the various agents (i.e., systems) whether they are humans or machines.

the composer), there will always be emerging properties popping up from its implementation that need to be managed in real-time (this is the role of the conductor). In addition, customers have to be involved in the vital aspects of the use of the product. For example, a commercial airplane is typically customized following specific requirements of an airline. Therefore, the main issue for the composer is to anticipate the best macro-structures and macro-functions that will allow for easy adaptation later on during customization.

Interaction among sub-systems is a key factor in the rationalization of SoS activity. From this point of view, the concept of system of systems tends to be very close to the concept of "society of agents" (Minsky 1985) developed in artificial intelligence and cognitive science. The cognitive function analysis approach was developed along these lines in human-centered design to enable analysis of highly automated systems in aeronautics (Boy 1998, 2011). The concept of function is inevitably embedded into the concept of an agent, and consequently into the concept of system of systems. In this book, we use the terms system-of-systems and multi-agent systems indifferently.

We saw that there are various types of interaction among agents (i.e., supervision, mediation and cooperation by mutual understanding). Therefore, human-centered design needs to be opened to various types of interactions among systems of systems (Landauer and Bellman 1996).

Analyzing SoS Properties

SoS properties are mainly related to **complexity**. Our worldwide economy has become virtual and less centralized. Therefore, more coordination and cooperation is needed for it to work. Outsourcing has also tremendously increased in order to decrease costs, generating many kinds of progressively standardized services. In the military domain, Command, Control, Computers and Communications Systems stress the need for more interoperability and synergism between systems. We usually talk about joint SoS, which could be associated with Hollnagel and Woods's joint cognitive systems (Hollnagel and Woods 2005).

Connectivity is certainly a main property of SoSs, especially because they are large and distributed. The Internet is a good example of such systems. Distribution is usually geographical and imposes a total virtual interaction among agents and systems (i.e., they exchange information instead of manipulating physical entities).

Emergent behavior is another important property of systems of systems, especially because their real functionalities emerge from various interactions that are enabled. The system performs functions and carries out purposes that do not reside in any component system. These behaviors are emergent properties of the entire SoS and cannot be localized to any component system. The principal purposes of the SoS are fulfilled by these behaviors. Education systems are also good examples of such systems.

Integration is an important process and property of SoS, especially because they are incrementally adjusted and adapted to improve synergy and cooperation. Integration requires strong technical coordination and cooperation (articulation work), as well as several kinds of appropriate specific expertise.

Operational and managerial independence enables each component to be used, fully operate and exist independently in case of disassembly of the SoS.

Evolutionary development is crucial because a SoS is never fully mature and complete (i.e., its functions are incrementally added, removed, modified, discovered and so on until maturity is satisfactorily achieved).

System-of-Systems Engineering (SoSE)

The methodology for defining, abstracting, modeling, and analyzing SoS problems is typically referred to as system-of-systems engineering (SoSE). Since a system of systems works through the collaboration of its components, the design of a SoS strongly requires modeling and simulation to master complexity, emergent behaviors, integration, operational and managerial independence, and evolutionary development. A good example is design and development of the Falcon 7X by Dassault, where an airplane was fully designed through the design of its components, simulating them independently and together generating emergent behaviors before any physical development was started.

SoSE is not so different from classical system engineering in terms of processes. The main difference is managing the incremental nature of the design and development of a SoS that requires a multi-agent model-based approach. Several attempts were taken to integrate SoSs into systems engineering such as the US Department of Defense guide for SoSE, and the inclusion of SoSs into the ISO/IEC 15288.[3]

Designing a new aircraft takes into account several disciplines (e.g., mechanical and aerospace engineering, electrical engineering, computer science and engineering, and human factors). Each agent collaborates with other agents to satisfy design needs and requirements. Such collaboration is central to SoSE. For example, flight tests should be coordinated with the design office in order to create feedback that fosters incremental modifications of various aircraft systems design. It is crucial that an architectural model be represented and used by various agents involved in the design and development of an aircraft (e.g., a large commercial aircraft involves a huge number of agents distributed in several countries that an organization needs to coordinate). Such an architectural model can be used as a mediating tool by all actors.

Discussion

Systems such as air-air surveillance capabilities (ADS-B) and cockpit automation (ASAS: Airborne Separation Assistance System) are being designed to enhance authority sharing between the flight deck and the ground. The evolution between what is currently done and the next generation of air-ground environments requires careful

---

[3] http://www.iso.org/iso/catalogue_detail.htm?csnumber = 43564.

scrutiny of function allocation and keeping automation as simple as possible, in terms of flexibility for the actors. Aircraft merging and spacing technology remains immature, requiring further investigation and development. In terminal areas, the merging and spacing process currently relies on air traffic controllers' skills and experience and is affected by weather conditions, rates of runway use, ground congestion and other factors. In the perspective of authority delegation to the flight deck, new approaches to merging and spacing need to be invented, especially in high-density traffic situations. These approaches will rely on new kinds of automated technology and procedures. Obviously, whenever merging and spacing can be anticipated en route, it would be a great gain of time and workload in terminal areas. It is now important to identify required functional evolutions and cognitive functions that emerge from this evolution, taking into account a representative environment with very high traffic. Referring to the Orchestra Model, new approach procedures and terminal area patterns are part of the common frame of reference (i.e., a music theory analog). Generic contracts, as scores, needs to be defined according to cognitive functions that will emerge from both new automation and organizational rules, mainly coordination rules. Contract coordination should be both anticipated (composer role) and managed (conductor role). Finally, function allocation should be thought in terms of authority sharing in the sense that several agents share responsibility and control in context. It could be a priori defined (i.e., each function represented by a contract is allocated to an appropriate agent). It should also be dynamically defined (i.e., cognitive function may be allocated with respect to an ongoing situation). As already seen, dynamic function allocation requires appropriate situation awareness (i.e., there is a constant need to look for potential hazards and understand perception and cognitive limits of various agents in order to compensate with additional cognitive functions and maintain an appropriate cognitive stability). Such cognitive functions could be additional resources in the form of supervisors, mediators or automated links that provide a better mutual understanding. Of course, their implementation and operational costs should be evaluated with respect to relevant human and technological factors. The choice of their effective implementation in the real world depends on these evaluations.

Other approaches, such as cognitive systems engineering /joint cognitive systems (Hollnagel and Woods 2005), consider the growing complexity of socio-technical systems, problems and failures of clumsy technology, and the limitations of linear models and the information-processing paradigm. They also recognize the need for cognitive function (Boy 1998) "in the mind" (i.e., processes that mediate responses to events). In fact, this anthropological approach of cognition was already started with the identification of situated actions (Suchman 1987) and distributed cognition (Hutchins 1995). All these contributions emphasize context as the main research issue. In fact, people are both goal-driven and event-driven; they are opportunistic according to context. This is why context is so important to identify and take into account. "Situated activity is not a kind of action, but the nature of animal interaction at all times, in contrast with most machines we know. This is not merely a claim that context is important, but what constitutes the context, how you categorize the world, *arises together* with processes that are coordinating physical activity. To be perceiving the world is to be acting in it–not in a linear input-output relation (act—observe—change)—but dialectically, so that what I am perceiving and how I am moving co-determine each other" (Clancey 1993).

Context is an extremely difficult subject to grasp and identify since it is directly associated to the persistence of situations and events (Boy 1998); some are long enough to be captured, and some others are too short to even be perceived. This is why a scenario-based approach carried out by domain-expert professionals is necessary. The Orchestra Model is a metaphoric model that enables handling context in a functional and structured way, since the cognitive function representation includes a context attribute by construction. Identification and categorization of the possible connections and interactions among agents through their cognitive functions enables us to better understand various relevant issues of situation awareness. In fact the way we identify and categorize the world is crucial in the perception of context when acting. It is clear that all metaphors are very limited, and the Orchestra metaphor has limitations when we use it to describe socio-technical systems. However, it incrementally emerged as an acceptable model of the evolution of our software-immersive multi-agent environment, and the ATM environment in particular.

Agent-oriented software engineering approaches are also developed (Jennings and Wooldridge 1998; Jennings 2000; Tveit 2001; Lin et al. 2006). Systems using software agents (i.e., Multi-Agent Systems or MAS's) are becoming more popular within the development mainstream because, as the name suggests, a cognitive agent aims to handle tasks autonomously with intelligence (i.e., a cognitive agent is both goal-driven and event-driven). To benefit from autonomous control and reduced running costs, system functions are performed automatically. Agent-oriented considerations are being readily accepted into various software design paradigms. Agents may work alone, but most commonly, they cooperate toward achieving some application goal(s). MAS's are components in systems that are viewed as many individuals living in a society and working together. Currently however, there is no universal agreement on how to build a comprehensive agent-oriented system. Development of MAS's is a non-trivial task especially without the necessary support provided by software engineering (SWE) environments. From a SWE perspective, solving a problem should encompass steps from problem realization, requirements analysis, to architecture design and implementation. These steps should be implemented within a life-cycle process and include testing, verification, and reengineering to prove that the built system is sound. Agent-oriented SWE techniques must be evaluated and compared to gain a better understanding of how agent systems should be engineered and evolved.

Even if agent-oriented software engineering only examines the software side of the problem of human and machine agent systems, it emphasizes the complexity issue by decomposing the overall system being designed into several sub-systems that are more manageable. This kind of approach is very similar to what we do by performing a cognitive function analysis of a complex human-machine system. In the same way, it is mandatory to abstract (i.e., deconstruct complexity) the resulting decomposed system in order to work on more salient aspects of the problem. The way this kind of abstraction is made is crucial since some "less relevant details" will be ignored. This is where human-centeredness is important. Once such an abstraction is performed, various kinds of relationships can be drawn between cognitive functions that determine the way an overall organization is supposed to work. Of course, this process has to be repeated until a satisfactory cognitive function network is found. It often requires human-in-the-loop simulations to find such a satisfactory cognitive function network.

As already described in a previous chapter (Boy 2002), cognitive function analysis has many similarities with **activity theory**, the Russian approach to cognition, which considers that people learn from their environment, and human activity is mediated by surrounding artifacts. The concept of cognitive function is very similar to Leont'ev's functional organs (Leont'ev 1981). "Functional organs are functionally integrated, goal-oriented configurations of internal and external resources. External tools support and complement natural human abilities in building up a more efficient system that can lead to higher accomplishments. For example, scissors elevate the human hand to an effective cutting organ, eyeglasses improve human vision, and notebooks enhance memory. The external tools integrated into functional organs are experienced as a property of the individual, while the same things not integrated into the structure of a functional organ (for example, during the early phases of learning how to use the tool) are conceived of as belonging to the outer world." (Kaptelinin 1995).

Another dimension that is not extensively presented is **time.** Time is very important in music. The Orchestra Model is a very insightful metaphor for time-wise investigations. We have already presented this by describing time sequences developed by the various cognitive functions involved in the Überlingen accident (Boy and Grote 2009). The specificity of the Orchestra Model is to encapsulate both design and performance times (i.e., the time of the composer and the time of the conductor and musicians). Information flows are important to capture in the form of useful and usable contracts (scores) designed and developed by composers at design time, and in the form of coordination patterns emerging from performance and handled by conductors at operations time.

Finally, agents interact among each other not only under supervision, through mediation, and by cooperation through mutual understanding; they may also compete. **Competition** is a model that was not taken into account here because we implicitly focused on systems where resources were sufficiently distributed among all agents. However, when for any reason there is a lack of resources, some agents may compete for the same available resources. In terms of cognitive functions, when resource space is sparse compared to the number of agents, conflicts start to emerge. For that matter, human-centered design needs to be started as a resource-based approach. A survey of various available resources should be made in the very beginning of a design process and incrementally upgraded as the design process evolves. For example, a resource could be a slot in an approach-and-landing sequence for an aircraft that may have to compete with another one. In the supervision model, the air traffic controller determines such resource allocation in real-time. In a mediation model where two aircraft have to interact with little knowledge about each other, there are risks if there is no mediator that understands the conflict and is able to solve it. In cooperation by mutual understanding, we could imagine that one aircraft could understand that the other has a higher priority since it does not have enough fuel for instance, and needs to land urgently.

Conclusion

Since context is a major concern in the design of appropriate life-critical systems, scenarios are very good tools to support the elicitation of emergent cognitive functions. Scenario-based design requires support by a strong conceptual model. The Orchestra Model is a good conceptual tool to categorize cognitive functions in air traffic management problems, their allocation among human and machine agents, as well as various relevant relationships between them.

The Orchestra Model defines relationships between agents, supported by contracts that are very similar to scores in music. In addition, when there are several agents to coordinate, these contracts (scores) need to be coordinated; this is typically the role of a composer in music. Despite initial planning (i.e., initial coordination of contracts), there are always events that are not anticipated either because they are intentions from agents that differ from the original plans, or unexpected external events. These events require dynamic re-allocation of functions, and therefore modification of initial contracts. This is typically the role of a conductor. Agents, as musicians, need not only to be competent in performing their functions; they also need to understand what the other agents are doing. This is why we also need interaction models, based on a common framework (music theory). In the best case, agents communicate between each other by mutual understanding, but they may require being supervised or mediated when they do not have acceptable situation awareness.

As many other contributors suggested, new technologies and automation do not have only quantitative effects, they also have qualitative shifts (Dekker and Woods 2005), induce the emergence of new practices (Flores et al. 1988), and even may alter the tasks for which they were designed (Carroll and Campbell 1988). The Orchestra Model provides a conceptual model that supports elicitation of these kinds of emergences (e.g., multi-agent activity patterns).

A question you may have after reading this chapter could be: How about having several orchestras performing at the same time? This is precisely what we are working on right now. Complex systems are typically systems of systems. as already presented, the concept of system-of-systems (SoS) is now widespread in the engineering of contemporary systems (Maier 1998; Lamb and Rhodes 2008; Sheard and Mostashari 2009). Current air traffic management systems, complex defense systems, the Internet, intelligent transport systems, and enterprise information networks (or Intranets) are systems of systems. The SoS concept tends to be very close to the concept of "agent", already described in this chapter as societies of agents. The cognitive function analysis approach was developed along these lines in human-centered design to enable analysis of software-intensive complex systems (Boy 1998). For that matter the concept of function is inevitably embedded into the concept of agent, and consequently the concept of system of systems (SoS). Today, building a SoS is a holistic endeavor that requires architects and appropriate tests to enable the emergence of salient properties and hidden cognitive functions. More generally, the concept of super-system emerges from the interaction among its components. In addition, there are various types of interaction among agents and systems of systems. Collaboration is one, but, as already discussed above, there are two others, like supervision and mediation, as well as competition. Therefore human-centered design should be opened to various types of interactions among systems of systems. For all these reasons, the Orchestra Model is interesting in many ways regarding support of SoS design and development.

References

Billings, C. E. (1991). *Human-centered aircraft automation philosophy*. NASA Ames Research Center, Moffett Field, CA, USA.

Boy, G. A. (1991). Advanced interaction media as a component of everyday life for the coming generation. Invited speech at he *JMA Marketing World Congress*, Tokyo.

Boy, G. A. (1998). *Cognitive function analysis*. Ablex. Westport: Greenwood Group. ISBN: 1-56750-377-2.

Boy, G. A. (2002). Theories of human cognition: To better understand the co-adaptation of people and technology, in knowledge management, organizational intelligence and learning, and complexity. In L. Douglas Kiel (Ed.), *Encyclopedia of life support systems* (EOLSS), Developed under the Auspices of the UNESCO. Oxford: Eolss. (http://www.eolss.net).

Boy, G. A. (2005). *Maturity, automation and user experience (Maturité, Automation et Experience des Utilisateurs)*. *Proceeding of the French Conference on Human-Computer Interaction*. New York: ACM.

Boy, G. A. (2007). Perceived complexity and cognitive stability in human-centered design. *Proceedings of the HCI International 2007 Conference*, Beijing.

Boy, G. A. (2009). The Orchestra: A conceptual model for function allocation and scenario-based engineering in multi-agent safety-critical systems. *Proceedings of the European Conference on Cognitive Ergonomics*. Finland: Otaniemi, Helsinki area, (30 September-2 October).

Boy, G. A. (2011). Cognitive function analysis in the design of human and machine multi-agent systems. In G. A. Boy (Ed.), *Handbook of human-machine interaction: A human-centered design approach*. Ashgate.

Boy, G. A., Salis, F., Figarol, S., Debernard, S., LeBlaye, P., & Straussberger, S. (2008). *PAUSA: Final technical report*. DGAC-DPAC, Paris.

Boy, G. A., & Grote, G. (2009). Authority in increasingly complex human and machine collaborative systems: Application to the future air traffic management construction. *In the Proceedings of the 2009 International Ergonomics Association World Congress*, Beijing.

Carroll, J. M., & Campbell, R. L. (1988). *Artifacts as psychological theories: the case of human-computer interaction*. IBM Research Report RC 13454, Watson research Center, Yorktown, Heights.

Carroll, J. M. (1995). *Scenario-based design: Envisioning work and technology in system development*. New York: Wiley.

Carroll, J. M. (2011). Scenario-based design. In G. A. Boy (Ed.), *Handbook of Human-Machine Interaction*, Ashgate.

Campbell, G., Cannon-Bowers, J., Glenn, F., Zachary, W., Laughery, R., & Klein, G. (1997). Dynamic function allocation in the SC-21 manning initiative program: Naval air warfare center training systems division, Orlando, SC-21/ONR S&T Manning Affordability Initiative.

Chapanis, A. (1965). On the allocation of functions between men and machines. *Occupational Psychology, 39,* 1–11.

Clancey, W. J. (1993). Situated action: A neuropsychological interpretation (Response to Vera and Simon). *Cognitive Science, 17*(1), 87–107.

Cockburn, A. (2001). *Writing effective use cases.* Boston: Addison-Wesley Longman. ISBN 0-201-70225-8.

Corso, G. M., & Moloney, M. M. (1996). Human performance, dynamic function allocation and transfer of training. In R. Koubek & W. Karwowski (Eds.), *manufacturing agility and hybrid automation-I. Proceedings of the 5th International Conference on Human Aspects of Advanced Manufacturing.* Louisville: IEA.

Dekker, S. W. A., & Woods, D. D. (2005). MABA-MABA or abracadabra? Progress on human-automation co-ordination. *Cognition, Technology & Work, 4,* 240–244.

Endsley, M. R., & Jones, W. M. (1997). *Situation awareness, information dominance, and information warfare* (No. AL/CF-TR-1997-0156). Wright-Patterson AFB: United States Air Force Armstrong Laboratory.

Faraj, S., & Sproull, L. (2000). Coordinating expertise in software development teams. *Management Science, 46*(12), 1554–1568.

Fitts, P. M. (Ed.). (1951). *Human engineering for an effective air navigation and traffic control system.* Washington, DC: National Research Council.

Flores, F., Graves, M., Hartfield, B., & Winograd, T. (1988). Computer systems and the design of organizational interaction. *ACM Transaction on Office Information Systems, 6,* 153–172.

Grande, M., & Visscher, M. B. (1967). *Claude Bernard and experimental medicine.* Cambridge: Schenkman.

Grote, G., Ryser, C., Waefler, T., Windischer, A., & Weik, S. (2000). KOMPASS: A method for complementary function allocation in auto-mated work systems. *International Journal of Human-Computer Studies, 52,* 267–287.

Hildebrandt, M., & Harrison, M. (2002). The temporal dimension of dynamic function allocation. *Proceedings of ECCE-11,* pp. 83–292.

Hildebrandt, M., & Harrison, M. (2003). Putting time (back) into dynamic function allocation. *Proceedings of the 47th Annual Meeting of the Human Factors and Ergonomics Society.* Denver, Colorado, October 13–17, pp. 488–492.

Hollnagel, E., & Woods, D. D. (2005). *Joint cognitive systems: Foundations of cognitive systems engineering.* Boca Raton: CRC-Taylor & Francis.

Hutchins, E. (1995). How a cockpit remembers its speeds. *Cognitive Science, 19,* 265–288.

Jacobson, I., Booch, G., & Rumbaugh, J. (1998). *The unified software development process.* Boston: Addison Wesley Longman. ISBN 0-201-57169-2.

Jennings, N. R. (2000). On agent-based software engineering. *Artificial Intelligence, Elsevier, 177,* 277–296.

Jennings, N. R., & Wooldridge, M. J. (1998). Applications of intelligent agents. In N. R. Jennings & M. J. Wooldridge (Eds.), *Agent Technology: Foundations, applications, and markets* (pp. 3–28). Heidelberg: Springer.

Kaber, D. B., & Endsley, M. R. (2003). The effects of level of automation and adaptive automation on human performance, situation awareness and workload in a dynamic control task. *Theoretical Issues in Ergonomics Science,* (pp. 1–40). Boca Raton: Taylor & Francis.

Kaptelinin, V. (1995). *Designing learning activity: a cultural-historical perspective in CSCL. Proceedings of the Computer Supported Cooperative Learning (CSCL'95).* Bloomington: Indiana University Press.

Kanki, B. G., & Foushee, H. C. (1989). Communication as group process mediator of aircrew performance. *Aviation, Space, and Environmental Medicine, 20*(2), 402–410.

Lamb, C. T., & Rhodes, D. H. (2008). Systems thinking as an emergent team property: Ongoing research into the enablers and barriers to team-level systems thinking. *SysCon 2008-IEEE International Systems Conference,* Montreal, Canada, April 7–10.

Landauer, C., & Bellman, K. L. (1996). *Collaborative system engineering and integration environments.* 5th International workshops on enabling technologies: Infrastructure for collaborative enterprises (WET ICE–96).

Leont'ev, A. (1981). *Problems of the development of the mind.* Moscow: Progress.

Lin, C. E., Kavi, K. M., Sheldon, F. T., Daley, K. M., & Abercrombie, R. K. (2006). A Methodology to Evaluate Agent Oriented Software Engineering Techniques. IEEE Proc. HICSS-40, Big Island HI, Software Agents and Semantic Web Technologies Minitrack.

Maier, M. W. (1998). Architecting Principles for Systems-of-Systems, *Systems Engineering, 1*(4), 267–284.

McCarthy, J. (1958). *Programs with common sense. symposium on mechanization of thought processes.* Teddington: National Physical Laboratory.

Malone, T., & Crowston, K. (1994). The Interdisciplinary study of coordination. *ACM Computing Surveys, 26*(1), 87–119.

Mathieu, J., Goodwin, G. F., Heffner, T. S., Salas, E., & Cannon-Bowers, J. A. (2000). The Influence of Shared Mental Models on Team Process and Performance. *Journal of Applied Psychology, 85*(2), 273–283.

Minsky, M. (1985). *The society of mind.* New York: Touchstone Simon and Schuster. ISBN 0-671-60740-5.

Newell, A., & Simon, H. A. (1972). *Human problem solving.* Englewood Cliffs: Prentice Hall

Parasuraman, R., Sheridan T., & Wickens, C. (2000). A model for types and levels of human interaction with automation. *IEEE Transactions on Systems, Man, and Cybernetics, 30,* 286–297.

Sheard, S. A., & Mostashari, A. (2009). Principles of complex systems for systems engineering. *Systems Engineering, 12*(4), 295–311, Winter.

Sheridan, T. B. (1987). Supervisory control. In G. Salvendy (Ed.), *Handbook of human factors* (pp. 1243–1286). New York: Wiley.

Straussberger, S., Chamayou, C., Pellerin, P., Serres, A., Salis, F., Feuerberg, B., Lantes, J. Y., Guiost, B., Figarol, S., Reuzeau, F., & Boy, G. A. (2008). *Scenarios in PAUSA.* DGAC-DPAC EURISCO, Technical Report. April.

Suchman, L. A. (1987). *Plans and situated actions: The problem of human-machine communication.* Cambridge: Cambridge Press.

Swain, A. D., & Guttman, H. E. (1980). *Handbook of human reliability analysis with emphasis on nuclear power plant applications* (NUREG/CR-1278) U.S. Nuclear Regulatory Commission, Washington, DC.

Thom, R. (1989). *Structural stability and morphogenesis: An outline of a general theory of models.* Reading: Addison-Wesley. ISBN 0-201-09419-3.

Tveit, A. (2001). *A survey of agent-oriented software engineering.* NTNU Computer Science Graduate Student Conf., Norwegian University of Science and Technology.

Ullman, M. T. (2004). Contributions of memory circuits to language: the declarative/procedural model. *Cognition, 92,* 231–270.

Walz, D. B., Elam, J. J., & Curtis, B. (1993). Inside a software design team: Knowledge acquisition, sharing, and integration, *Communications of the ACM, 36*(10), 63–77.

Weyer, J. (2006). Modes of governance of hybrid systems. The mid-air collision at Ueberlingen and the impact of smart technology. *Science, Technology & Innovation Studies, 2.* ISSN: 1861-3675

Wickens, C. D. (1992). *Engineering psychology and human performance* (2nd edn.). New York: Harper Collins. ISBN: 0673461610

Wright, P. C., Fields, R. E., & Harrison, M. D. (2000). Analyzing human-machine interaction as distributed cognition: The resources model. *Human-Computer Interaction, 15*(1), 1–4.

Chapter 3
Cognitive Engineering

Introduction

Recently, Norman (2011) wrote, "Machines have rules they follow. They are designed and programmed by people, mostly engineers and programmers, with logic and precision. As a result, they are often designed by technically trained people who are far more concentrated about the welfare of their machines than the welfare of the people who will use them. The logic of machines is imposed on people, human beings who do not work by the same rules of logic." Isn't it obvious? Nevertheless, this is what we observe everyday, and very little is being done in engineering to solve this recurring problem effectively. This kind of observation has been made for a long time by ergonomists who preached the adaptation of machines to people and not the opposite. What is new is the consideration of this requirement not as a post-development validation of machines (i.e., human factors and ergonomics, or HFE), but as a pre-design process, as well as a life cycle iterative process (i.e., human-centered design, or HCD). **Cognitive engineering** is about understanding people's needs and experience along the life cycle of a product, and most importantly with influence during its high-level requirements definition.

The AZF accident on September 21st, 2001, in Toulouse, France, is an unfortunate dramatic example of a rare event that was hardly difficult to predict, but was certainly possible to prevent if there would have been more care. Here, a distinction must be made between prediction and prevention. As already mentioned in the introduction of this book, prediction is necessarily valid in the short term. Similar to the effect of smoking on the development of cancer: you cannot predict it, but you certainly can prevent it by envisioning possible results. In the Toulouse accident, 300 tons of ammonium nitrate (NH_4NO_3) exploded, destroying the whole factory. The blast measured 3.4 on the Richter scale, with an estimated power equivalent to 20–40 tons of trinitrotoluene (TNT). The whole factory was destroyed making a crater of depth 20–30 m (65–100 feet deep), with a diameter of 200 m (650 feet); steel girders were found 3 km away from the explosion (Barthelemy et al. 2001). There were 30 fatalities and an estimated 10,000 injuries. Can we say that this kind of accident was easy to predict? At least 16 major ammonium nitrate

G. A. Boy, *Orchestrating Human-Centered Design*,
DOI 10.1007/978-1-4471-4339-0_3, © Springer-Verlag London 2013

disasters[1] had already occurred prior to the Toulouse accident in 2001. Therefore, we cannot say this is an exceptional accident. Since 1947 following the Texas City disaster, US regulations for storage and handling were established. Since then we know that large stockpiles of the material can be a major fire risk and may detonate due to their supporting oxidation. Experience feedback is crucial, but it appears that we tend to forget to apply prevention after long periods of "**routine success**". We become complacent. The Fire and Blast Information Group (FABIG) has provided the following lessons learned (FABIG 2011) that illustrate the re-discovery of experience feedback from 1947:

- Redefine ammonium nitrate to cover lower percentage composition.
- Improve quality of hazard studies and their homogeneity between different industrial sites. Studies should specify basic assumptions concerning accident scenarios, external threats and failure of safety systems.
- Reduce risks posed by hazardous installations via various measures such as double confinement and breaking up stock into smaller amounts.
- Harmonize regulation requirements for transport of goods in areas such as ports and marshalling yards.
- Define new land use planning rules that deal with potential hazardous situations.

I remember when the AZF accident happened. I was in Toulouse attending a meeting not too far from the accident location, and clearly heard the detonation. My mother, who was a teenager at the time when the AZF factory was built (fertilizer production began at that plant called ONIA in 1927) and lived close to it, told me that for many years the population had to go through evacuation exercises, a practice that came to an end after the second world war. Even if this would not have had an effect in the 2001 accident, this is an additional fact that confirms complacency and the killing effect of routine success. Note also that land around the factory was much less populated in 1927 than in 2001. The main point is, **engagement** and **prevention** are real assets in the management of life-critical systems.

Since prediction is impossible in the long term, we need to establish goals, and manage to achieve them. In life-critical systems, these goals are commonly based on safety, efficiency and comfort. **Long-term anticipation** is necessarily goal-driven fueled by creativity, competence, knowledge and excellent management. Otherwise, when people do not have the sense and courage to plan long-term, taking the necessary means to reach established goals, these individuals are simply reacting to events in the short term. People usually persist in this kind of short-term event-driven behavior because most events are manageable, and therefore reactive behavior is acceptable until an **unexpected event** occurs. However, when events are not manageable—we call them accidents—normal reactions are transformed into crisis management. Let me then distinguish between long-term goal-driven anticipation and short-term event-driven activity. Human-centered design necessarily belongs to the former distinction, even if it should take into account the latter.

---

[1] http://en.wikipedia.org/wiki/Ammonium_nitrate_disasters.

In the current short-term event-driven economy, our technology-centered society generates products that are not fully mature and integrated. A product is mature when it is ready to use (for the kiwi fruit, we would say "ripe"), and safe (for the kiwi fruit, we would say "healthy"). **Maturity** is a matter of care (i.e., high-level requirements must be well-established and valid, with iterative tests using appropriate principles and criteria. The maturity of a product is necessarily related to the maturity of practice (e.g., people who do not care for healthy food cannot see the difference between a properly ripened fruit and another product processed for business reasons only). Sometimes criteria are different among people (e.g., quantity may be more important than quality and vice versa).

It is common to observe that we tend to accumulate independently useful artifacts without considering their **integrated effect**. For example, car equipment (e.g., radio, global positioning system, phone hand-free kits, collision avoidance systems and speed control) are great individually, but when they are put together, they may become a nightmare for the driver who needs to manage them. Multitasking was extensively analyzed in aeronautics (Loukopoulos et al. 2009). It becomes a key phenomenon. Why? It is quite clear that the lack of integration is the reason. Technology-centered approaches tend to promote individual products because they are easily sellable by suppliers. Manufacturers should be more proactive and design in a human-centered manner.

We cannot describe what is going on today without talking about our increasingly software-immersed world and its exponentially growing complexity, even if the original intentions of engineers were to simplify it. We added layers and layers of software artifacts where nothing existed before. Sometimes it is great! Sometimes it is not so great! In this book, human-centered design will often refer to software-intensive systems, and more specifically **interactive systems**.

The Famous "Human Error" and Automation

For almost three decades the field of human factors and ergonomics has focused on "human errors" and "human reliability" (Norman 1988; Reason 1990; Hollnagel 1993, 1998; Sarter and Woods 1995; Hollnagel and Amalberti 2001). When I started working at ONERA (the French Aerospace Lab) in 1980, my job was to develop an HFE approach for the certification of the newly designed two-crewmen cockpits for commercial aircraft. At that time, people were interested in **reducing the number of human errors** because the Aviation Safety Reporting System (ASRS), started in 1976, announced that over 70 % of reports (aviation incidents) contained evidence of human error (Billings and Cheaney 1981). For that matter, automation was seen as a means to avoid human errors. This engineering logic was based on the observation that if some functions were devoted to the machine instead of people, they would be performed according to specifications. What we did not see at that time was that people adapt to technology no matter what. Surprisingly, they adapted to automation. . . and new types of human errors occurred! In addition, automation was mostly implemented because technology was available, but the difficult question of "what function for whom or what" was not properly addressed. Yet, function allocation has been studied for a long time (Fitts 1951), and is not an a priori concept and process. **Function allocation** is dynamic in the sense that there are **emergent functions** emerging from practice (i.e., from interactions among human and machine agents). The term "machine agent" is used in this book to denote "automation", mostly represented by software systems.

In the early eighties, HFE was evolving from health and safety medicine to experimental psychology. Cognitive engineering was nascent. HFE specialists were using theories and concepts from past experience. A technological revolution was growing and Airbus Industrie was the first commercial aircraft manufacturer to design and deliver highly-automated cockpits for commercial airplanes. The question was, "how will we certify these new highly-automated machines?" HFE disciplines did not have appropriate concepts and tools to respond effectively. It is interesting to observe that "cognitive engineering" was growing in parallel as a major discipline capable of supporting such an endeavor. Donald Norman was the first to coin the term "cognitive engineering" in 1982. During this period, many scientists put down the basis for cognitive engineering research such as Sheridan, Rasmussen, Hollnagel, Woods, Winograd, Suchman and Hutchins. Cognitive engineering was born within the growing evolution of cognitive science, cognitive psychology, artificial intelligence, computer science, electrical and computer engineering, and software engineering. Several new disciplines emerged such as human-computer interaction, computer graphics, speech recognition, natural language generation, mobile technologies, ubiquitous computing and, more generally, advanced interaction media.

Automation is a good solution for improving safety when it is mature and done well (Boy 2011). What really changes is the nature of practices, and more specifically human errors. People may commit errors no matter what, "*errare humanum est!*" The main problem is to understand their nature and master their repercussions. We know that control of industrial complex systems is typically supported by operational procedures. Such **procedures** are a kind of automation (i.e., **automation of people behavior**). This distinction between human automation and machine automation will be further developed in Chap. 4. When these procedures are translated into software, they become machine automation. Procedural interfaces are possible solutions for this kind of translation (Boy 2002). In fact, human automation can be supported by various characteristics. First, training and development of skills are traditional solutions. Competent people are people who train and prepare themselves for specific complex tasks. If we use the model of Rasmussen (1983), this refers to the development of a skill-based behavior, typically used sub-consciously. Second, people may learn procedures and rules that are consciously used at Rasmussen's rule-based behavior level. Sometimes, people need external explicit support at this behavioral level. This support is mainly composed of written procedures. Third, people may need to use highly cognitive resources (i.e., Rasmussen's knowledge-based behavior level). At this level, they need to identify the situation that is often either unanticipated or unknown, decide among a set of alternative response solutions, plan the chosen response solution and act. Of course, at any of these three levels, there may be cooperation between people and machines; one supports the other and conversely, and this is where function allocation takes place.

At this point, let us go back to **human errors and machine failures**. In such a multi-agent environment, we want each agent to support the other in case of an error or failure (i.e., human-machine cooperation). We will see later in this book that we are talking about hyper-redundancy. Indeed, **redundancy** is a good solution for anticipating errors and failures, and eventually minimizing repercussions when they occur. There are human errors that are risky (i.e., repercussions can be tremendous). It is clear that we want to resist to this type of errors. **Error resistance** means that we do not want these errors to have any impact on the stability of the overall human-machine system. However, there are other types of errors that do not lead to serious repercussions, and we can tolerate them. **Error tolerance** can be a good strategy because people can learn from such errors in an immunological way. These two perspectives, i.e. error resistance and tolerance, have been heavily studied during the last two decades and depart from the initial view of reducing the number of human errors (Billings 1996; Reason 1990).

Like in physics, **stability** in human-machine systems can be:

- passive (i.e., a system that deviates from its stable state is able to return to this state passively (by itself)), or
- active (i.e., the system requires active assistance to return to its stable state when disturbed).

The concept of resilience is sometimes preferred to the concept of stability because it represents the ability of a system to absorb shocks and reorganize while retaining its essential structure and identity. In the same conceptual neighborhood, the concept of homeostasis is also relevant since it captures the regulation of internal stability of a system. Overall, I choose stability in human-machine systems as a central concept, because it is related to turbulence, control, dynamics, equilibrium, balance and fault (in the sense of error and failure).

The right to make errors can be very beneficial in the development of a safety culture. Indeed, risk taking is often necessary when no predefined solution is available, and people need to be able to make mistakes without being punished. Note that in life-critical systems, these people are usually experts in their field. The main issue is not over-precaution, but prevention, **competence** and good judgment. The problem today is that finance-driven management too often finds expert operators too expensive; they prefer cheaper solutions. In addition, note that competence is not only a matter of people and technology, but also a matter of organizational structure.

From Purposes to Means: Inside-Out Versus Outside-In

Technology-centered engineering typically goes from **means to purposes**. It follows an approach that goes **from inside-out**[2] (i.e., first we engineer a technology and then try to find out the uses of it). The builder/engineer metaphor is applicable here in the

---

[2] I recently discussed with a young architect who told me that this term could be confusing since in architecture, architects have to perceive, understand and project themselves inside the house

Fig. 3.1 Engineering
builder's view. (Copyright·
ClipartOf.com)

sense that a builder or an engineer puts technology first and so on until an artifact is
built, and eventually tries to find out how to use it (Fig. 3.1).

Human-centered design goes from **purposes to means**. It follows an approach
that goes **from outside-in** (i.e., first we try to find out what would be the various
uses of the artifact to be built, and define the appropriate technology to achieve the
usability goals). The architect /designer metaphor is applicable here in the sense that
if you want to build a house and call an architect /designer to help you, he or she will
ask you how you usually live, if you have a family, if you receive guests often, if you
have children and so on. . . and will create a mock-up that you will be able to evaluate
from your own perspective. This user requirements gathering supports the creation
of a first mock-up that can be presented and evaluated by the clients. Mock-ups are
incrementally refined to satisfy evolving user requirements. Indeed, user requirements are very likely to evolve because clients adjust their needs to technological
possibilities (i.e., they are usually not aware of technological possibilities before they
see them). Note that the term "client" may be misleading when an architect talks
with a company. Clients can be top managers or other employees. It is important to
find out who the real users are, as well as how these users currently interact and will
interact among each other. This is why the outside-in approach to design requires

---

(interior) they are building to investigate human factors issues reflecting them in the exterior and
structure as well. In this book, "inside-out" refers to the inside of technology, the kernel, bricks,
beams, engines and so on. It means that design is done from the kernel of technology to the periphery
where users are (i.e., taking a technology-centered approach). In fact, when I talk about architects,
I mean human-centered designers, the ones who try to find out how houses they build will be used.
In addition, some architecture schools currently tend to move toward engineering. This is why
human-centered design should have its role as a discipline in its own right.

Fig. 3.2 Participatory design scene

interaction models that support this kind of analysis. We already provided three types of models of interaction among agents: supervision; mediation and cooperation by mutual understanding. Looking for interaction models is a good HCD approach that can be carried out using incremental functional analysis and participatory design (Fig. 3.2).

In the inside-out approach, human factors are usually taken into account when the product is fully developed. The main quest is to develop a user interface and related operational procedures. This is a commonly used approach today. User interface specialists try to adjust people to the machine that has already been built.

In the outside-in approach, human factors are taken into account during early stages of the design process and all along the product life cycle. The main goal is to find out how to best allocate functions between users and product being designed. Why in some sectors, such as architecture and design, is this approach very well mastered, and why in some others, such as software and system engineering do we still need to make progress? An answer to this question could be that architecture exists as a discipline for several millennia and is mature; on the other hand, systems engineering is still very young. Design thinking comes today as both a first step, chronologically, and a necessary glue for architecture and engineering.

What Is Cognitive Engineering?

Cognitive engineering is a composite discipline. It certainly deals with engineering and computer science, but it also encompasses human and social sciences. More specifically, cognitive engineering is typically well served by disciplines such as cognitive science and artificial intelligence, human-computer interaction, human factors and ergonomics, as well as anthropology and organization sciences. Cognitive engineering takes into account human capabilities and limitations in the design of systems. Typical cognitive human factors are workload, fatigue, human errors, performance, risk taking, decision-making and situation awareness. Cognitive engineering progressively became an inescapable discipline for humans, as computers introduced and established dominant cognitive work as opposed to physical work. In addition, we needed to have workable methods to explain why trained and expert human operators commit coarse errors. An important goal of cognitive engineering is to render socio-technical systems more reliable through cognitive modeling.

There is **cognition in engineering** because for the last three decades technology has incorporated a tremendous amount of software to the extent that any system built today includes a growing amount of information technology. We have put layers and layers of software in life-critical systems to improve safety, efficiency and comfort. We have incrementally created unprecedented complexity. As an example, the latest aircraft cockpits are being equipped with data-link technology for communication. This technology provides a massive amount of data that needs to be processed before presented to pilots. Conventional technology is still VHF radio systems on many aircraft, which is much more simple (i.e., VHF complexity is very well understood), but in the early eighties the ASRS showed that radio communication was the main cause of information transfer problems (Billings and Cheaney 1981). They showed that 85 % of information transfer problems came from audio (i.e., radio, interphone, voice, telephone, tape recording), and 15 % from visual (i.e., video CRT, instruments, lights, publications, charts).

In addition, it is important to mention the nature of these information transfer problems included message not originated (37 %), inaccurate, incomplete, ambiguous and garbled (37 %), correct but untimely (13 %), not received or misunderstood (11 %), and finally not transferred due to equipment failure (3 %). Billings and Cheaney's 1981 NASA report concluded "these data make it quite apparent that verbal communication was an imperfect method of information transfer. For that reason, much attention has been given to alternative means of transferring the information that is now communicated almost exclusively by voice. Much effort was devoted to the design of digital data link via satellite. This solution removes most of the problems such as noise and sequential nature of spoken messages. The first data-link systems were coded-text based. The second generation of data-link systems is based on the observation that communication is mainly related to navigation activities, therefore graphical displays are being designed to show the navigation scene around the aircraft (e.g., cockpit display of traffic information (CDTI)), 30 years after Billings and Cheaney's 1981 report. CDTI is not even in service. What is the difference between VHF and CDTI? Integration! Integration of communication and navigation requires understanding the various cognitive patterns involved in navigation and air traffic management activities. This exemplifies cognitive engineering (i.e., finding such cognitive patterns). They could range from taking a different flight level to spacing and merging during an approach and landing phase. Understanding these patterns does not mean only enumerating them, it also means finding the **emerging cognitive functions**[3] that are induced by the automation of such patterns. Again, automation

---

[3] What are emerging cognitive functions? First, let us define a human cognitive function, which can be modeled by the role they play in the organism where they are involved. It should be noted that

could be either represented by a set of procedures that pilots have to perform, or machine automation on CDTI displays, as an example. In the latter case, there can be several levels of automation (Sheridan and Verplank 1978).

There is **engineering in cognition** because even if cognitive science was far from engineering, some difficult real-world problems now have solutions that involve understanding human and machine cognition as a whole. Rasmussen's model is a perfect example of a model built by an engineer, and further used by the most prominent psychologists. Cognitive science and psychology need such models to support their studies and explain experimental results. Human factors should be taken into account during the whole life cycle of a product and mainly during the first stages of a design. How can we anticipate users' cognitive functions during design? By definition, we cannot analytically anticipate emerging cognitive functions. We need to run human-in-the-loop simulations to discover and identify such functions or properties. They emerge from interactions among various agents.

In addition, designers should be aware of such emergence. First, they need to understand the concept of emerging cognitive function. Second, they need to understand that without appropriate **scenarios and criteria**, they will not be able to discover emerging cognitive functions. Third, they need to be proactive and **take reasonable risks** to provoke the emergence of such functions. For that matter design teams should be multidisciplinary, innovative and curious. Fourth, I strongly believe that designers and engineers should be trained in a **human-centered perspective**. Designing the right thing from the beginning is key. Consequently, designers' cognition is a research topic in itself that should be better investigated. This is also why I prefer the term "human-centered design" to "user-centered design", since there are many stakeholders (i.e., humans, dealing with products to be designed, developed, used, maintained and eventually decommissioned). End-users constitute one category of people.

Design Should be Holistic and Deeper at the Same Time

The design of a life-critical system is hard because it needs to address possible failures and consequences from various perspectives that should be both broad and deep at the same time. This is difficult. I strongly believe that successful complex systems

---

this role is defined in a limited context, which also needs to be defined. Finally, a cognitive function is necessarily supported by resources, which can be cognitive functions themselves. Consequently, a cognitive function can be defined by the triplet, "role, context, resources" (Boy 1998, 2011). Note that both humans and information-based machines have cognitive functions. For example, people have situation awareness and decision-making cognitive functions; speed control systems installed on cars have regulation mechanism that enable to maintain a given speed, and collision avoidance systems enable aircraft to inform pilots that they are approaching another flying vehicle and they need to immediately take action to avoid collision. When designing a new system, cognitive functions can be defined deliberatively as prescriptions, but they most naturally emerge from human activity. Emerging cognitive functions cannot be easily anticipated at design time. Furthermore, it may take some time to discover them. This is why tests are so crucial (e.g., flight testing in aeronautics).

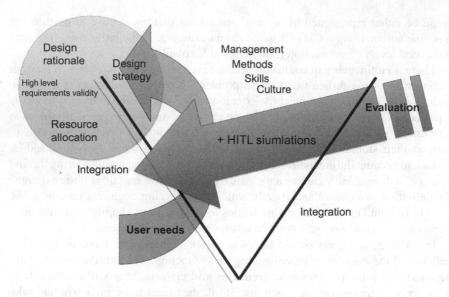

Fig. 3.3 Human-centered design V-model of the life cycle of a system

are built by knowledgeable, skilled, motivated and well-coordinated people. These people should have the same high-level goal (e.g., going to the Moon). Quality of the overall vision is essential.

Many industries use the systems engineering **V-Model** to deconstruct the complexity of the system development cycle. The left branch of the V, devoted to project definition, includes the concept of operations, requirements and architecture, detailed design, and leads to implementation. The right branch of the V, devoted to project test and integration, includes integration, test, verification, validation, operations and maintenance. This model is very much technology-centered. It can become human-centered if we add the following properties and processes (Fig. 3.3):

- early design ideas should be fed by **users' needs**;
- **design rationale** should be clearly defined with respect to purpose;
- **high-level requirements** should be well defined with a continuously tested function allocation between the various agents (humans and machines);
- **integration** is not a second step process, it should be a first step process (i.e., the project definition should be holistic); and
- **human-in-the-loop simulation** should be used very early to inform and define high-level requirements. HITLS enables designers and engineers to discover, identify and validate emerging cognitive functions as much as possible. HITLS is also useful during the entire life cycle of the system.

It is time to provide more details on the way human-centered design supports integration of a complex human-machine system. First, the human-machine system is an organized set of human-machine sub-systems resulting in a system of systems

Fig. 3.4 A system as a
network of sub-systems

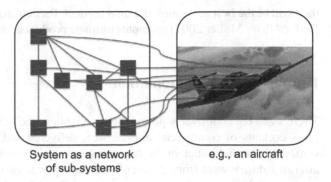

System as a network
of sub-systems

e.g., an aircraft

Fig. 3.5 From the cognitive
architecture of a
human-machine system to the
socio-cognitive organization
to build and manage it

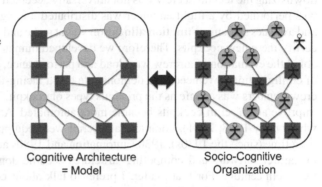

Cognitive Architecture
= Model

Socio-Cognitive
Organization

(Fig. 3.4). As already seen, each human-machine sub-system can be represented by
a network of human and machine cognitive functions.

The overall interconnected cognitive function network is usually called the **cognitive architecture**. Integration of the overall human-machine system results from
the right allocation and coordination of people and teams to the appropriate cognitive functions of the cognitive architecture (Fig. 3.5). I strongly suggest using the
AUTOS[4] pyramid (Boy 2011) to support the life cycle of a human-machine system
from design to decommissioning. Integration and maturity of such a system can be
accomplished by using and comparing successive AUTOS pyramid models as design
and development processes evolve.

Of course, this approach can be handled and supported by various kinds of appropriate use case representations (Alexander 2001; Cockburn 2001; Jacobsen 1992;
Kulak and Guiney 2000; Lilly 1999). Both analysis and description of use cases are
good because they enable the designer to discover points that are not clear during
the requirement specification process. We usually start from vague requirements that
need to be better defined and more importantly validated before it is too late. End
users have to be involved in the development of use cases not only because they can
incorporate their needs and desires in the design of a new system, but also because

---

[4] Artifact, User, Task, Organization and Situation. The AUTOS pyramid will be more extensively
described in Chap. 7—devoted to Modeling and Simulation.

they will be the best advocates of a final product. Participatory design (Grudin 1993; Bødker 1996; Muller 2007) promotes ultimate product acceptance at delivery.

Human-Centered Automation

There was a big controversy during the late eighties when the first highly-automated glass cockpits of commercial aircraft were delivered and used. This controversy started with social issues in the beginning of the eighties because the commercial aircraft industry went from three-crewmen cockpits to two-crewmen cockpits, and downsizing the technical crew was not universally accepted. The role that previously was performed by a flight engineer was distributed among the captain, first officer and avionics systems. **Function allocation** was at stake and we had to find out how to certify these new cockpits. Therefore we developed human factors methods that enabled the evaluation of aircrew workload and performance, comparing various types of configurations. We needed to demonstrate that a facing-forward cockpit with two crewmembers was as safe as the previous types of cockpit. Technology continued to improve and aircraft cockpits became more automated. As already described elsewhere (Boy 1998, 2011), commercial aircraft were equipped with autopilots for a long time (since the 1930s). 1930s automation and 1980s automation were different in nature; software and computing structures became dominant in the end of the twentieth century. For that matter, I prefer to talk about **computarization** instead of automation. We actually started to incorporate more software into cockpits. The very notion of systems quickly became persistent and pilots' work radically changed from handling flight qualities (manual control) to aircraft **systems management**. A pilot's role shifted from control to management, exactly like when someone becomes a manager in an organization and has a team of agents to manage. In this case, pilots had to learn how to manage very advanced systems and coordinate their activities. It was not obvious when suddenly a pilot had to become a manager (of systems). This new emerging cognitive function (i.e., systems management), had to be learned and stabilized.

Situation Awareness, Decision-Making and Risk Taking

Being aware of your environment when you are controlling and managing a life-critical system is a crucial issue. Many researchers and practitioners have developed fundamental work on situation awareness (Beringer and Hancock 1989; Billings 1995; Endsley 1988, 1995, 1996). In aviation for example, the aircrew needs to be aware of external situations (e.g., weather conditions, traffic, terrain) as well as internal situations (e.g., systems states, failures, fuel level).

Why is **situation awareness** (SA) so crucial? Again, the never-ending integration of software into systems not only creates a bigger distance between people and physical machines, but also the very nature of human-machine interaction changed and continues to change. Software enables the emergence of new interaction styles and experience. The main issue is to provide the right interaction capability at the right time and in the right way. This is difficult. Why? The situation awareness concept includes perception of the situation, its comprehension and the necessary means of projection in order to act safely, efficiently and comfortably. Having the right machine affordances and useful skills to be learned supports SA.

To be more specific, flying an airplane requires constant situation understanding in order to perceive and anticipate upcoming events. For fuel management, the pilot must be aware of the state of the fuel level in the selected tank; must understand the rate variation of fuel use, detect abnormalities and reorganize the way engines are fueled; and must determine if he or she is unlikely to reach destination due to the unexpected rate of fuel use and divert to an alternate airport prior to engine failure (Spirkovska 2010). We see in this example that pilots need to have the right level of information at the right time to decide and act appropriately. Designers have to find solutions that take into account limited amounts of physical space for presenting the situation and potentially provide appropriate guidance. Usability engineering (Nielsen 1993) and more generally human factors and ergonomics (Meister 1999) provides methods to this end, but without creativity and expert continuous testing such complex systems cannot be properly designed.

Situation awareness is not only individual, it can be distributed among a set of agents whether humans or artifacts (Artman and Garbis 1998). Referring to an Orchestra model, we have seen that authority can be shared and distributed (see Chap. 2). For that matter, situation awareness and decision-making can be shared and distributed. Distributed situation awareness is therefore "an emergent property of collaborative systems, something that resides in the interaction between elements of the system and not in the heads of individual operators working in that system." (Salmon et al. 2009)

Mathematicians have tried to model **decision-making** by using many kinds of approaches such as expected utility (Bernoulli 1738) and later on game theory (Von Neumann and Morgenstern 1944). Obviously, decision-making cannot be reduced to a pure cognitive mechanism, usually seen as an algorithmic process, which enables the analysis of all possible alternatives and leads to the choice of the best one. Decision-making can be described with respect to various dimensions such as rational versus irrational, cognitive versus emotional, goal-driven versus event-driven. In life-critical environments, it entails risk taking. In all cases, decision-making involves three main attributes:

- goals and objectives;
- alternatives; and
- selection principles, criteria and processes.

At this point, we need to make a distinction between **normative or prescriptive** models and **descriptive** models. The former tell what people should do, the latter tell what people actually do when facing a decision. In their seminal papers on prospect theory, an alternative model to expected utility theory as a descriptive

Fig. 3.6 The SRAR model

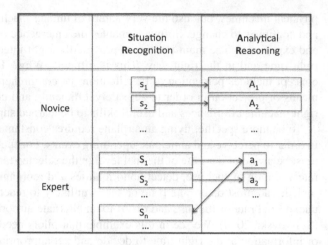

model of decision-making under risk, Kahneman and Tversky (1979) claimed that "people underweight outcomes that are merely probable in comparison with outcomes that are obtained with certainty." They called this tendency, the **certainty effect**, which "contributes to risk aversion in choices involving sure gains and to risk seeking in choices involving sure losses." Kahneman and Tversky replaced probabilities by decision weights on gains and losses, and these weights are generally lower than the corresponding probabilities, except in the range of low probabilities. This is interesting in the case of ultra-safe systems (Amalberti 2001), such as nuclear power plants or aviation systems, where probabilities of failure are extremely low. More generally, people make decision—using heuristics learned from experience (Wickens 1987; Mosier-O'Neill 1989; Evans 1990). We will come back on this issue when we develop the abduction inference mechanism.

Dreyfus and Dreyfus (1980) brought up the distinction between novice and expert users during the rise of expert systems in artificial intelligence. This is related to skill set, knowledge and motivation that someone has in accomplishing a task using a system. I carried out a study at NASA involving both novices and experts in the manipulation of a specific system of the space shuttle called the orbital refueling system (ORS). Both novices and experts were pilots, with a difference that experts were acquainted with ORS operations (Boy 1986, 1987). From these studies, I deduced the **Situation Recognition and Analytical Reasoning** (SRAR) model presented in Fig. 3.6.

SRAR states that novices have small situation patterns used to recognize a situation, usually provided from designers' instructions (e.g., procedures), which lead to complex analytical reasoning algorithms (i.e., these patterns are usually the result of failure mode analyses performed by engineering services (design offices)). SRAR also states that experts develop and use sophisticated situation patterns, typically constructed through learning and experience, which lead to very fast analytical reasoning actions. In fact, these patterns induce familiarity with specific situations, and their richness provides a far better mental model of the situation and related appropriate actions. A learning model was deduced and presented in (Boy 1998). Experts' situation patterns are typically preconditions produced from chunking (Rosenbloom and Newell 1987). Note that unlike the novices, when an action of the analytical reasoning does not provide an acceptable solution, experts may go back to situation recognition to investigate a new solution (arrow going backward from a_1 on Fig. 3.4). They use both forward and backward chaining inference mechanisms (i.e., they could be qualified as educated opportunists). This opportunistic behavior is usually very fast, and consequently experts may see more than one backtracking from analytical reasoning to situation recognition to solve a problem.

These early models anticipated what is now called Naturalistic Decision Making (NDM). Indeed, more recent attempts to better understand human decision-making in work situations were developed by Klein (1997) and the NDM community. Klein and Klinger (1991) proposed ten features of NDM:

- ill-defined goals and ill-structured tasks;
- uncertainty, ambiguity, and missing data;
- shifting and competing goals;
- dynamic and continually changing conditions;
- action-feedback loops (real-time reactions to changed conditions); time stress;
- high stakes;
- multiple players;
- organizational goals and norms;
- experienced decision makers.

These approaches to decision-making are grounded in the **situated cognition** movement started by Vygotsky, Barlett and Dewey's work in the early 1900s in reaction to cognitive work that tended to ignore context. During the 1970s, the Chilean school of biology provided a substantial account to situated cognition by defining autopiesis (Maturana and Varela 1980). Suchman brought situated action into play in cognitive engineering (Suchman 1987), and Sperber developed the concept of massive modularity that provides the mind with flexibility and context-sensitivity (Sperber 2005).

Decision leads to action and, in life-critical systems, inevitably to **risk taking**. It is then important to take into account the type of risk-taker who will make the decision. When an individual takes a risk, there exist some elements, which by definition are not completely under control. Minimizing risk involves reducing to a minimum both the number of these elements and their level of danger.

According to Llewellyn (2003) who studied rock climbing activities, there are three different risk-taking types: the "risk avoiders" who avoid activities due to the risks involved; the "risk reducers" who participate in high-risk activities in spite of the risks involved; and the "risk optimizers" who participate in high-risk activities partly because of the risks involved (Fig. 3.7).

The decision making process involved is mainly that of abduction, one of the three forms of logical inference, along with deduction and induction. It consists of making the hypothesis of a consequence and then proving the inference $(A \rightarrow B)$ leading from a premise A to the consequence B. The mental process of abduction, notably

Fig. 3.7 Llewellyn's possible relationship between level of perceived risk, risk orientation and risk acceptability. (Adapted from Llewellyn 2003, p. 27)

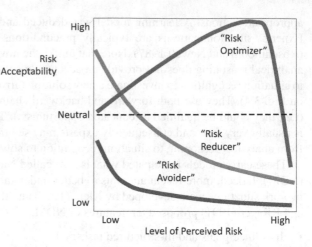

the choice of the correct hypothesis, is closely linked to pragmatism, intuition, expertise and skill. Charles Peirce defines **abduction** as the process of constructing an explicative hypothesis and maintains that it is the only logical operation by which to introduce a new idea (Peirce 1958). Airbus's fly-by-wire electronic controls were revolutionary in transport aircraft. The designers at the time set up a process of abduction. Of course, this choice gave rise to a number of deductive processes, which contributed to rationalizing and proving the pertinence and viability of the system. For instance, one might seek to deduce how to make an aircraft lighter, facilitate operator training, protect against external undesirable events or human error. Initial abductive processes, and the rational deduction processes which follow are guided by general objectives related, for example, to operational economics, safety, performance, comfort and training. It is the state of knowledge at a given moment, which provides the basis for the choice and implementation of these processes. The definitive choice of fly-by-wire controls resulted in fact from reasoned and deductive choices with regard to specific objectives such as weight reduction, simplicity, advantages of protecting the flight envelope and homogeneity of the different training courses.

Risk taking is an everyday process in which we are involved from a young age. When a baby tries to walk, he or she is taking risk (without knowing of course), and this process is necessary to figure out what walking really means in practice. Help does not hurt, but the baby has to discover the various states and movements that entail walking until he or she is really walking. Automatisms are incrementally assimilated and accommodated (in Piaget's sense), embodied and forgotten from a cognitive standpoint (i.e., they become subconscious). Later on in our lives, we take risks for several reasons such as improving performance, rescuing other people in danger, and looking for pleasure. Risk taking requires accumulated trial and error experience, an incremental learning from this experience. At this point, we need to make a distinction between voluntary and involuntary risk taking. Voluntary risk taking is goal-driven (i.e., intentional); involuntary risk taking is event-driven (i.e., reactive).

Anytime we voluntarily take risk, we postulate a possible future that we will have to demonstrate (i.e., we use abductive inference). Mountaineers who try to climb the Mount Everest are experts, and prepare themselves physically, psychologically, and organizationally. These adventurers should not start before they are fully ready. All abductive inferences they will perform along to way to the summit will be guided by heuristics acquired during their preparation and experience.

Involuntary risk is not such a different matter. Anytime we are facing a crisis to manage, it is always better to have appropriate abductive inference mechanisms that were prepared prior to the crisis. More generally, crisis management should be based on preparation and experience. Pilots train in full-flight simulators to deal with very rare situations such as engine failure, fuel leaks, and depressurization. They construct responses to these rare events in the form of automatisms (skills), rules and knowledge. Of course, there are other events that are more unexpected, and sometimes unknown in advance. These events require different kinds of abilities such as creativity, boldness, concentration, calm, charisma in addition to mental availability and rapidity. Risk takers know that they always need to identify recovery areas, or backups, where they can stabilize in case of external aggression. External aggression is any event that comes to destabilize the current way of doing things.

During the risk taking conference that I co-organized in 2008, Manu Gaidet, a triple world champion of ski freeride, explained how he went about his preparation: "When looking at a slope, I imagine my descent, the "line" I want to follow. I analyze the paths of possible avalanches or snowfalls.... I choose the itinerary according to an estimation of the speed, my physical condition at the time and my level of self-confidence. I identify "zero risk" areas and dangerous paths. And then I imagine how to go from one zero risk spot to another... My golden rule is: "if I can't picture it, I don't do it!" (Boy and Brachet 2008).

Risk management is about avoiding known risks and preparing for crisis management. The former has been addressed by many programs that frame risks in industry, and is strongly operationalized. The later remains unfortunately weak. Crisis management requires inter-disciplinary skills and knowledge that have to be combined by mentally-stable leaders, such as astronauts. The more situations and responses that can be anticipated, the better, obviously. Knowledge of past crises and how they were managed is useful experience that needs to be capitalized upon. Therefore, crisis managers need to have long-term training and experience, as well as situation awareness, decision-making and planning capabilities. Finally, risk taking cannot be successful without strong motivation.

From Command and Control to Knowledge Work

Rasmussen's model is useful to explain the evolution of work during the last century or so (Rasmussen 1983). In a few words, the human performance model, developed by Jens Rasmussen and extensively used in human factors and ergonomics, is

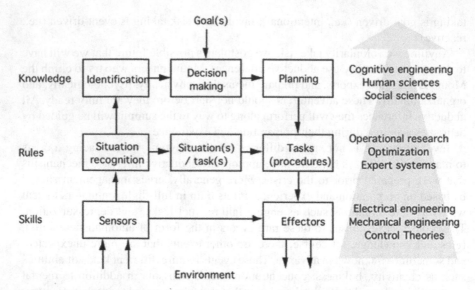

Fig. 3.8 Rasmussen's model, automation and contributing disciplines (Boy, 1998)

articulated around three levels of human behavior: skills, rules and knowledge (Fig. 3.8). At the skill-based level, people develop sets of perception-action skills that they commonly use subconsciously. This level of behavior was transferred to machines since the beginning of the twentieth century (e.g., steam machine regulators, washing machines, airplane autopilots (since the 1930s on commercial airplanes), and more recently cruise control on cars). Several engineering disciplines provide automation solutions for the transfer of human skills to machines (e.g., electrical engineering, mechanical engineering and control theories).

At the rule-based level, people develop production rules in the form of "if this situation occurs then use this task". Rules are commonly used consciously. This level of behavior was transferred to machines during the eighties (e.g., flight management systems, onboard commercial airplanes, performing navigation rules that pilots used to perform). Again, engineering and computer science contributed to provide solutions for the transfer of human rules to machines (e.g., operations research, optimization and expert systems).

At the knowledge-based level, people develop highly cognitive functions such as situation identification (i.e., they need to make sense of a perceived problematic situation that was not recognized by the rule-based level), decision-making (i.e., they need to investigate a set of possible alternatives and choose one of them in relation with their overall goals), and planning (i.e., they need to plan a set of articulated tasks that would solve the problem). To date there is no machine that can handle this knowledge level. Consequently, cognitive engineering and more generally human and social sciences have a great role to play at this level, and are contributing to provide solutions.

Work is currently involving more knowledge. This phenomenon grew during the last few decades (Davenport 2005). Workers can be compared more to musicians than to the soldiers of the past. Today's workers need to be able to perform at the knowledge level. Of course, this is possible within a specific subject area. Sometimes we talk about subject matter experts or knowledge workers. The concept of "knowledge workers" has still to be defined more precisely despite several attempts (Mosco and McKercher 2007). The art of **knowledge work** involves creativity, design thinking, as well as information searching, processing and exchange. Social media are now an unavoidable support to knowledge work. Knowledge work is shaping the twenty-first century. It goes with mobility and ubiquitous computing (see description in Chap. 8 on advanced interaction media). Complexity arises from the evolution of our information-intensive world where interconnectivity provides both structure and functions. Work has moved from doing to thinking.

If we refer to Rasmussen's model, work in the twentieth century was very much centered on rules and procedures stored in books and binders (i.e., paper based). In many cases, these rules and procedures were learned by human operators, but paper support was used to help human operators use checklists and do-lists during operations. We converged toward rule-based behavior that is currently preferred to skills and knowledge because it is cheaper and actually works in most situations. However, there are new situations (almost unpredictable) that lead to catastrophic consequences. In these situations, we need technology and organizations that support both skill-based behavior (e.g., facilitating intuitive thinking, informational navigation and manipulation), and knowledge-based behavior (e.g., facilitating comprehension and projection).

Discussion and Conclusion

Several times, I have heard two "popular" claims about cognitive engineering: (1) cognitive engineering is systems engineering with human factors in it, and (2) cognitive engineering is a specific branch of human factors and ergonomics (HFE).

The former claim is typically provided by systems engineers who consider that people can be easily reduced to boxes as specific systems among technological systems. People are certainly complex, highly nonlinear systems (e.g., people usually have several solutions to a problem, but you never know which one they will finally choose). What people can do as "systems" is difficult, and most of time impossible, to predict. Think about complex systems that lead to singularities, bifurcations, and attractors (Kellert 1993). Of course, when they are very well trained, experienced and follow well-established procedures, their predictability can be very high. However, despite high levels of expertise and experience, it is well known that more than 75 % of incidents and accidents in aviation are due to human errors. This is why many human factors studies have been developed during the last 3 decades in this domain. But, it is also known that most of human errors are recovered, and most importantly qualified human operators successfully solve problems that no machine could do. Therefore, even if we may think people could be the problem in systems-engineering-with-human-factors-in-it, they can effectively be the solution when the human-system integration is done well (i.e., when appropriate cognitive engineering principles, methods and tools are effectively used from the very beginning of the design process).

The latter claim generally comes from the current HFE community, which I consider old fashion from many points of view. First and foremost, HFE developed as a discipline that was, and still is, used to evaluate systems already designed and developed. Consequently, it contributed to adapt people to machines, even if the HFE credo is the opposite. Of course, a clear opposition emerged between engineers who designed and developed systems, and human factors specialists who tried to adapt these systems to people by providing guidelines for the development of user interfaces and procedures/user-guides. Second and more recently, HFE is now recognized as an important part of engineering from a regulatory standpoint (i.e., many standards have been developed to the point that HFE is very similar to quality assurance). The main problem with this approach is that since industrial HFE is now reduced to procedure following and intense reporting, we tend to lose highly qualified specialists who are replaced by form filling operators.

In this chapter, we have seen that cognitive engineering emphasizes the way high-level requirements are designed by taking into account user needs and requirements, and by placing great emphasis on testing. For that matter, the use of modeling and simulation methods and tools is crucial during the whole life cycle of a product. Cognitive engineering deliberately takes people first, and is an **outside-in** approach to engineering. Systems engineering deliberately takes systems first, and is an **inside-out** approach to engineering. Therefore these two disciplines are fundamentally different by nature.

It is interesting to observe that people seem to place more attention on aircraft and nuclear power plants (NPP) accidents than on car accidents, despite the fact that the number of deaths due to car accidents is tremendously higher than aircraft and NPP accidents. Why? From a cognitive engineering point of view, the question of authority is at stake (i.e., who is in charge). When you drive a car you are in charge of driving or most of the time you know the person who is driving. Commercial aviation and the nuclear industry are managed and controlled by people who you generally do not know but need to trust. The main issue is that you have no possible impact on control of the life-critical system you are in.

We are back to the TOP model, where we are constantly facing the emergence of new practices and still need to ensure socio-cognitive stability of the organization as we are developing new technology. Today, an important concern is speed. Technology innovation and production advance much faster than before. Various kinds of problems are in the forefront such as interoperability of systems, technology and practice maturities, difficulty of adaptation and resistance to changes. It is now important that we improve understanding of our socio-technical evolution.

References

Alexander, I. (2001). Visualizing requirements in UML. *Telelogic Newsbyte*, Issue 13, Sept–Oct, http://www.telelogic.com/newsbyte/article.cfm?id=0000201800032832. Accessed 15 April 2012.

Amalberti, R. (2001). The paradoxes of almost totally safe transportation systems. *Safety Science, 37*, 109–126.

Artman, H., & Garbis, C. (1998). Situation Awareness as Distributed Cognition. Proceedings of ECCE '98, Limerick.

Barthelemy, F., Hornus, H., Roussot, J., Hufschmitt, J. P., & Raffoux, J.F. (2001). Accident on the 21st of September 2001 at a factory belonging to the Grande Paroisse Company in Toulouse. Report of the General Inspectorate for the Environment, October.

Beringer, D. B., & Hancock, P. A. (1989). Exploring situational awareness: A review and the effects of stress on rectilinear normalization. In *Proceedings of the Fifth International Symposium on Aviation Psychology* (Vol. 2, pp. 646–651). Columbus: Ohio State University.

Bernoulli, D. (1738). Exposition of a new theory of measurement of risk. *Econometrica* (The Econometric Society trans: Dr. L. Sommer, 1954 Vol. 22 (1), pp. 22–36).

Billings, C. E., & Cheaney, E. (1981). *Information transfer problems in the aviation system*. NASA TP-1875. Moffett Field: NASA Ames Research Center.

Billings, C. E. (1995). Situation awareness measurement and analysis: A commentary. *Proceedings of the International Conference on Experimental Analysis and Measurement of Situation Awareness*. Florida: Embry-Riddle Aeronautical University Press.

Billings, C. E. (1996). *Aviation automation: The search for a human-centered approach*. Mahwah: Erlbaum.

Bødker, S. (1996). Creating conditions for participation: Conflicts and resources in systems design. *Human Computer Interaction, 11*(3), 215–236.

Boy, G. A. (1986). An expert system for fault diagnosis in orbital refueling operations. *AIAA 24th Aerospace Sciences Meeting*, Reno.

Boy, G. A. (1987). Operator assistant systems. International journal of man-machine studies. In G. Mancini, D. D. Woods & E. Hollnagel (Eds.), *Cognitive engineering in dynamic worlds* (Vol. 27, pp. 541–554). London: Academic.

Boy, G. A. (1998). *Cognitive function analysis*. Ablex: Greenwood. ISBN 9781567503777.

Boy, G. A. (2002). Theories of human cognition: To better understand the co-adaptation of people and technology, in knowledge management, organizational intelligence and learning, and complexity. In L. D. Kiel (Ed.), *Encyclopedia of life support systems* (EOLSS), developed under the auspices of the UNESCO. Oxford: Eolss. http://www.eolss.net.

Boy, G. A. (2002). Procedural interfaces (in French). *Proceedings of the National Conference on Human-Computer Interaction (AFIHM)*. New York: ACM.

Boy, G. A. (2011). Cognitive function analysis in the design of human and machine multi-agent systems. In G. A. Boy (Ed.), *Handbook of human-machine interaction: A Human-centered design approach*. Aldershot: Ashgate.

Boy, G. A., & Brachet, G. (2008). *Risk taking. Dossier of the air and space academy*. Toulouse: ASA.

Cockburn, A. (2001). *Writing effective use cases*. Addison-Wesley.

Davenport, T. H. (2005). *Thinking for a living: How to get better performance and results from knowledge workers*. Boston: Harvard Business School Press. ISBN 1591394236.

Disasters caused by ammonium nitrate (2010). http://en.wikipedia.org/wiki/Ammonium_nitrate_disasters. Accessed 16 Dec 2011.

Dreyfus, S. E., & Dreyfus, H. L. (1980). *A five-stage model of the mental activities involved in directed skill acquisition*. Operations Research Center, ORC-80-2. Berkeley: University of California.

Endsley, M. R. (1988). *Situation awareness global assessment technique (SAGAT)*. Paper presented at the National Aerospace and Electronic Conference (NAECON), Dayton.

Endsley, M. R. (1995). Measurement of situation awareness in dynamic systems. *Human Factors,* *37,* 65–84.

Endsley, M. R. (1996). Automation and situation awareness. In R. Parasuraman & M. Mouloua (Eds.), *Automation and human performance: Theory and applications* (pp. 163–181). Mahwah: Laurence Erlbaum.

Evans, J. (1990). *Bias in human reasoning: Causes and consequences.* London: Lawrence Erlbaum Associates.

FABIG (2011). *Major accident listing: AZF (Azote de France) fertilizer factory.* Toulouse. March. http://www.fabig.com/Accidents/AZF + Toulouse.htm. Accessed 17 Dec 2011.

Fitts, P. M. (1951). *Human engineering for an effective air navigation and traffic control system.* Washington DC: National Research Council, Committee on Aviation Psychology.

Grudin, J. (1993). Obstacles to participatory design in large product development organizations. In A. Namioka & D. Schuler (Eds.), *Participatory design. Principles and practices* (pp. 99–122). Hillsdale: Lawrence Erlbaum Associates.

Hollnagel, E. (1993). *Human reliability analysis: Context and control.* London: Academic.

Hollnagel, E. (1998). *Cognitive reliability and error analysis method: CREAM.* New York: Elsevier.

Hollnagel, E., & Amalberti, R. (2001). The emperor's new clothes, or whatever happened to "human error"? Invited keynote presentation at 4th International Workshop on Human Error, Safety and System Development. Linköping.

Jacobsen, I. (1992). *Object oriented software engineering: A use case driven approach.* Addison-Wesley.

Kahneman, D., & Tversky, A. (1979). *Prospect theory: An analysis of decision under risk. Econometrica* (Vol. 47, No. 2, pp. 263–292 March).

Kellert, S. H. (1993). *In the wake of chaos: Unpredictable order in dynamical systems.* Chicago: University of Chicago Press. ISBN 0-226-42976-8.

Klein, G. A. (1997). The recognition-primed decision (RDP) model: Looking back, looking forward. In C. E. Zsambok & G. A. Klein (Eds.), *Naturalistic decision-making.* Mahwah: Lawrence Erlbaum Associates

Klein, G. A., & Klinger, D. (1991). Naturalistic decision-making. *Human Systems IAC Gateway* (Vol. XI: No. 3, 2, 1, pp. 16–19) Winter.

Kulak, D., & Guiney, E. (2000). *Use cases: Requirements in context.* Addison-Wesley.

Lilly, S. (1999). Use case pitfalls: Top 10 problems from real projects using use cases, proceedings of the technology of object-oriented languages and systems, IEEE.

Llewellyn, D. J. (2003). The psychology of risk taking behavior. PhD Thesis, The University of Strathclyde.

Loukopoulos, L. D., Dismukes, R. K., & Barshi, I. (2009). *The multitasking myth—Handling complexity in real-world operations.* Aldershot: Ashgate. ISBN: 978-0-7546-7382-8.

Maturana, H. R., & Varela, F. J. (1980). Autopoiesis: The organization of the living. In H.R. Maturana & F.J. Varela (Eds.), *Autopoiesis and cognition: The realization of the living* (pp. 59–138). Dordrecht: D. Reidel.

Meister, D. (1999). *The history of human factors and ergonomics.* Mahwah: Lawrence Erlbaum Associates. ISBN 0805827692.

Mosco, V. & McKercher, C. (2007). Introduction: Theorizing knowledge labor and the information society. In C. McKercher, V. Mosco & M. D. Lanham (Eds.), *Knowledge Workers in the Information Society* (pp. vii–xxiv). Lexington Books.

Mosier-O.Neill, K. L. (1989). A contextual analysis of pilot decision-making. In R. S. Jensen (Ed), *Proceedings of the Fifth International Symposium of Aviation Psychology.* Columbus: Ohio State University

Muller, M. J. (2007). Participatory design: The third space in HCI (revised). In J. Jacko & A. Sears (Eds.), *Handbook of HCI* (2nd ed). Mahway: Erlbaum.

Nielsen, J. (1993). *Usability engineering.* Boston: Academic Press.

Norman, D. (1988). *The psychology of everyday things.* New York: Basic Books.

Norman, D. A. (2011). *Living with complexity.* Cambridge: MIT.

Peirce, C. S. (1958). *Science and philosophy: Collected papers of Charles S. Peirce* (Vol. 7). Cambridge: Harvard University Press.

Rasmussen, J. (1983). Skills, rules, knowledge; signals, signs and symbols, and other distinctions in human performance models. *IEEE Transactions on Systems, Man and Cybernetics, 13,* 257–266.

Reason, J. (1990). *Human error.* Cambridge: University Press.

Rosenbloom, P. S., & Newell, A. (1987). Learning by chunking, a production system model of practice. In D. Klahr, P. Langley, R. Neches (Eds.). *Production system models of learning and development* (pp. 221–286). Cambridge: MIT.

Salmon, P. M., Stanton, N. A., Walker, G. H., & Jenkins, D. P. (2009). *Distributed situation awareness: Theory, measurement and application to teamwork.* Aldershot: Ashgate. ISBN: 978-0-7546-7058-2.

Sarter, N. B., & Woods, D. D. (1995). How in the world did we ever get into that mode? Mode error and awareness in supervisory control. *Human Factors, 37*(1), 5–19.

Sheridan, T. B., & Verplank, W. (1978). *Human and computer control of undersea teleoperators.* Cambridge: Man-Machine Systems Laboratory, Department of Mechanical Engineering, MIT.

Sperber, D. (2005). Modularity and relevance: How can a massively modular mind be flexible and context-sensitive? In P. Carruthers, S. Laurence & S. Stich (Eds.), *The innate mind: Structure and content.* Oxford: Oxford University Press.

Spirkovska, L. (2010). Intelligent automation approach for improving pilot Situational awareness. NASA Ames Research Center.

Suchman, L. (1987). *Plans and situated actions: The problem of human-machine communication.* Cambridge: Cambridge University Press.

Von Neumann, J., & Morgenstern, O. (1944). *Theory of games and economic behavior* (2nd ed. 1947, 3rd ed. 1953). Princeton: Princeton University Press.

Wickens, C. D. (1987). Information processing, decision-making and cognition. In G. Salvendy (Ed.), *Handbook of human factors.* New York: Wiley.

Chapter 4
Life-Critical Systems

Introduction

Aircraft, spacecraft, nuclear power plants (NPP's), operating rooms in hospitals and hospitals themselves are usually denoted as "safety-critical systems". They also belong to a broader class called "life-critical systems" that covers the purpose of this book and beyond. Life-critical systems (LCS) can be very simple such as circuit breakers, fire alarms, or fuse. They also can be very sophisticated such as robotic surgery machines, NPP instrumentation and control (I & C) systems, scuba-diving equipment, railway signaling and control systems, advanced braking systems on cars, air traffic management systems, and spacecraft.

Someone asked me, "How do you teach life-critical systems?" Well, the field is progressing and we have more knowledge and knowhow than 30 years ago for sure! In fact, we have more anecdotes and stories to tell than formal theoretical knowledge. Each domain has developed its own **expertise**, and some people have developed more expertise than others in the same domain. Therefore, it is time to collect this expertise, even in the form of a storytelling experience, massage it, compare various stories, and try to discover **emergent properties**. During the last 3 years, I have developed and gave the Life-Critical Systems course at Florida Institute of Technology in a variety of domains. Specialists from various life-critical domains have given lectures[1]; they were engineers, scientists and operational people, who shared with my students strong technical and/or operational experience on this kind of LCS.

A life-critical system may have one, two or three of the following major dimensions: **safety**, **efficiency** and **comfort**. Imagine that you lose your smart phone. What would happen? You would start to think about what you have stored in it, your contacts, your agenda including notes about some people you know, and so on. You would think about the danger induced by a malicious use of this data. In addition,

---

[1] These lectures were given at the FIT Human-Centered Design Institute by Thierry Bellet, John Hansman, Ludovic Loine, Patrick Millot, Amy Pritchett, Antoine Rauzy, and Robert Wears. I asked my students to analyze these contributions and find out commonalities and differences, make a synthesis and report on life-critical systems. Students found this approach very interesting and insightful, and most of them did not hesitate to go deeper into theoretical LCS knowledge. Final reports were excellent.

you would almost immediately get another device of that type because the need created by your lost smart phone would persist. Smart phones are life-critical from these perspectives; they involve safety, efficiency and comfort.

LCSs are characterized by several properties. They must be reliable, (i.e., tolerant or resistant to human error and system failure) resilient, redundant, socio-cognitively stable, as well as controllable and observable. They require operator's training. They also induce the generation of appropriate operating procedures, specific human skills, human operator selection, regulations and heavy (more standardized) design and development processes.

Again, **software** is everywhere. It is in cell phones, cars, public places, homes, and appliances of every kind including computer-aided design (CAD) tools that designers use to produce clothes today. Such systems are not only used as tools; they are also able to interact with people. If we focus on the observation that computers are disappearing inside appliances that fit people's needs and lives (Norman 1998), human-centered design is an evolution of the field of human-computer interaction, which has emerged during the last 3 decades.

As we have seen in the previous chapter, agents may fail: human agents may make errors, and machine agents may have system failures. In general, such **failures** may be the cause of serious consequences, and unfortunately sometimes-catastrophic accidents. This is why LCSs need to be taken seriously, and analyzed, designed and tested with respect to appropriate scenarios, criteria and principles.

The Need for Domain Expertise and Situated Cognition

This brings us to the difficult problem of **expertise**. Many contributions have been produced since the glorious time of expert systems and knowledge engineering (LaFrance 1987; Boose and Gaines 1988, 1990; Bradshaw et al. 1993; Hoffman et al. 1995). What can these contributions bring to the HCD field today? Situated at the intersection of artificial intelligence, cognitive psychology and philosophy, even if these contributions did not address HCD directly, introduced a very valuable framework. Other contributions came from the HCI field (Sutcliffe 1997), anthropology (Hutchins 1995), plus human engineering and human factors (Vicente 1997).

All these theories are great, but they will never eliminate the involvement of **domain experts** in the design process. You cannot design an aircraft if you do not have both aerospace engineering and operational aviation practice knowledge and knowhow. Complex operations have to be handled by expert people. This cockpit in Fig. 4.1 looks tremendously complex; it is the Concorde cockpit with about 600 instruments for three crewmen. Twenty Concorde aircraft were built (more concretely, 14 airline aircraft), pilots learned how to use them and did not have any accident with it in 27 years of operations, except the Paris accident in 2000 that was caused by an external unpredictable mechanical event. Concorde pilots were experts and they were able to manage the complexity of this aircraft routinely. We need to mention that to become a Concorde pilot, they had to train and qualify through the entire

Fig. 4.1 Experimental test Pilot André Turcat in the Concorde cockpit. (Courtesy of the Air and Space Academy, photo: Sud-Aviation/ collection Sparaco)

chain of aircraft first, i.e. knowledge gained by experience was also considered as a requirement. Analogously, climbing the Everest requires climbing many other smaller mountains prior to the Everest climb.

Complexity, expertise and life-critical systems are intimately related. I am regularly told by some of my friendly visitors that they do not understand how I can manage with my desk, with many layers of documents, reports, books and post-its allover the place. It looks like a mess from the outside, but believe me I know where things are. Why? This is a matter of implicit **situated and cognitive indexing**. Sometimes, documents are placed on top of each other chronologically. Some other times, they are organized by topics or proximity to my current task. There are various kinds of organizations that look invisible, are not understandable by outsiders, but are extremely effective in the everyday life. In reality, the outsider's perceived complexity of my desk turns out to be a very organized internal complexity in my brain. . . well; and nobody can access my thoughts, except me! I must admit that I get lost from time to time, just because I saturate my own cognitive indexing capacity. This is why I also use post-it notes to upgrade this implicit indexing of the documents on my desk by an explicit one. This is certainly a concrete example demonstrating the way we observe and that the world is highly cognitive, depending on our own knowledge, knowhow and experience. This is why it is so important to construct cognitive representations and models that help us observe and manage life-critical systems in a more educated way.

In addition, I also noticed that by increasing the size of my desk I did not reduced its perceived complexity. Indeed, I observed that I even put more documents on it. The desktop is a **cognitive attractor** (Lahlou 2000, 2005). This concept is also valid for the metaphor of the desktop on our contemporary computers. It is so easy to put files on the desktop that we usually do it! The entropy[2] keeps increasing until it is not acceptable any longer, and we have to clean it. I recently experienced very bizarre phenomena on my computer; for some reason, it started to get slower and slower, and suddenly restarted at a normal speed. These phenomena were random. I initially thought that this was due to Internet and external attacks. But after I turned off my WIFI capability, I realized that the same phenomena re-appeared. It took me a while before I understood that my computer **memory** was almost full. This is something I experienced 30 years ago when our memory capacities were minimal. But I forgot this as the years progressed, since we now seem to have almost infinite memory capacity. Most of the problem was coming from my emails that I keep religiously, just in case I would need them! In practice, I never need them, especially most of those that were sent several years ago... except when one day I exceptionally need one of them. This example shows that since our human memory can remember things that happened many years ago, we tend to replicate the same pattern on our machines. As our socio-cognitive world is evolving, we start to perceive emergent behaviors and properties of that kind.

In the previous example, complexity was coming from the number of emails and finally the limited capacity of my computer, but also from my own perception of the "infinity" of computer memory. Capacity limitation introduced a non-linearity or a discontinuity in my own linear perception of the management of my emails. Think about ATM today and the constant increase of aircraft in the sky. The sky is wide, but airport airspace vicinities have limited capacities. Increasing these capacities is a very complex problem from an ATM point of view.

This chapter will provide insights and grounds for comparing various kinds of life-critical systems. Human-centered design of life-critical systems is difficult. We need metrics to categorize **emerging properties** of such systems. Appropriate organizational set-ups need to be put in place. Technology is critical. Legal and regulatory requirements need to be addressed and developed, for certification in particular. For example, life-critical avionics (i.e., electronic systems on aircraft) is produced using procedures based on standards that are incrementally upgraded.

Major Life-Critical System Dimensions

Three major life-critical systems dimensions are presented: safety, efficiency and comfort.

---

[2] Here, entropy refers to disorder, complexity and finally chaos.

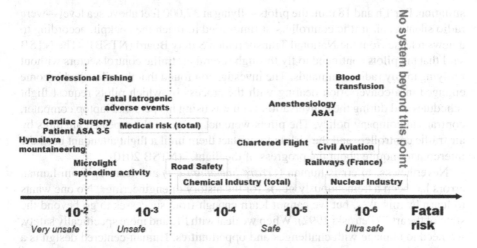

Fig. 4.2 Amalberti's risk exposure scale of safety-critical systems. (Courtesy of René Amalberti)

Safety

Many authors have already described the safety aspect of life-critical systems, such as Amalberti et al. (2005) in the aviation domain first and then in the medicine domain. Amalberti's risk exposure scale (Fig. 4.2) provides a very meaningful description of various domains with respect to levels of risk. For example, Amalberti calls **ultra-safe systems**, systems that have a rate of catastrophic accident exposure, such as complete failure of an aircraft engine leading to hull loss, which is better than one death per million exposures (10^{-6}). Conversely, for very unsafe domains, such as mountaineering or cardiac surgery, professional risk awareness is very high.

The term "professional risk awareness" is used because there can be, in contrast, unprofessional or unconscious risk awareness when people do not know what they are doing. I will not develop this part of the problem in this book, which focuses on the quest for expertise and experience. Being an astronaut cannot be improvised; it takes a long training and experience to perform such activity. The same is true for the cardiac surgeon who develops a skill set and knowledge on cardiac surgery for many long years before being able to perform his or her job autonomously and safely.

What is interesting to notice is that the more we get into higher levels of safety, the more people lose risk awareness. People become more confident because the system almost always works safely. Several phenomena may then emerge, such as **complacency** and drowning into excess of procedures and rules. When routine becomes a habit and everything goes well, people tend to be more complacent and release constraints. Such behavior may result in undesirable consequences. For example, on October 21, 2009, two Northwest airline pilots missed their airport and kept flying for 150 miles before landing the plane and its 147 passengers safely, because they were in a passionate discussion over airline policy and they lost awareness of the situation. For 1 h and 18 min, the pilots—flying at 37,000 feet above sea level—were radio silent as air traffic controllers at times tried to reach the cockpit, according to a news release from the National Transportation Safety Board (NTSB). "The NTSB said that the pilots continued to fly through several air traffic control sectors without replying to any radio commands. The investigation found that the pilots had become engaged in a conversation dealing with the process by which pilots request flight schedules and during the conversation each was using his personal laptop computer, contrary to company policy. The pilots were not aware of the repeated attempts by air traffic controllers and the airline to contact them until a flight attendant used the intercom to inquire about the progress of the flight." (NTSB 2010).

Nevertheless, to err is human (*errare humanum est*) and learning from human errors has been a tremendously guide for the history of engineering. "No one wants to learn by mistakes, but we cannot learn enough from successes to go beyond the state of the art." (Petroski 1992). When we deal with life and more specifically safety, we need to manage with challenges and opportunities. Human-centered design is a matter of function allocation, which is extremely difficult a priori and requires lots of testing (i.e., errors and corrections along the line). Function allocation is always a matter of situation awareness, decision-making and risk taking.

We have seen in Chap. 2 that function allocation is not a new issue. The MABAMABA approach (Fitts 1951) describing the respective strengths of people and machines has been very much used and even criticized by some human factors specialists. This approach is based on the assumption that functions are allocated a priori, but we know that there are **emergent cognitive functions** that cannot be discovered without experience feedback. Human-centered design should then be used to better articulate these emergent cognitive functions and appropriate automation and software. The key solution here is to create experience. Modeling and simulation is obviously a solution at design time as during the whole life cycle of a product.

Efficiency

Efficiency deals with availability, maturity, effectiveness, sustainability and ease of recovery after failure. The cost of inefficient systems is a criterion for the rejection of these systems. Reliability of people and machines can always be improved.

There are several possible models for representing system availability. **Availability** can be modeled as the ratio of the available time of a functional unit to the sum of the available time and the unavailable time of this functional unit. Availability can also be defined as the probability of a system to be available at a given time. Availability is also related to system reliability (Pagés and Gondran 1986; Lewis 1996; Der Kiureghian et al. 2005). System reliability has been studied for a long time and there are many techniques and solutions available today. **Redundancy** is key in life-critical systems. An LCS could continue to operate when part of it fails; we usually say that it is fail-passive and manual reversion should be possible; e.g., autopilots on

Table 4.1 ESA's technology readiness levels summary. (Data is in the public domain, copyright free)

TRL	Level description
1	Basic principles observed and reported
2	Technology concept and/or application formulated
3	Analytical and experimental critical function and/or characteristic proof-of-concept
4	Component and/or breadboard validation in laboratory environment
5	Component and/or breadboard validation in relevant environment
6	System/subsystem model or prototype demonstration in a relevant environment (ground or space)
7	System prototype demonstration in a space environment
8	Actual system completed and "Flight qualified" through test and demonstration (ground or space)
9	Actual system "Flight proven" through successful mission operations

airplanes. An LCS could also become safe when they cannot operate; we usually say that it is fail-safe (e.g., a medical infusion pump or an electronic door that can automatically unlock itself in case of failure). This fail-safe approach can be opposed to a fail-secure approach (e.g., electronic doors that lock when power failure) where systems maintain maximum security when they cannot operate (Wikipedia 2011). Finally, some LCSs can be fault-tolerant such as the built-in test equipments (BITE) on commercial airplanes.

Of course, efficiency cannot be expected before the system becomes mature (i.e., system's technology becomes mature and its associated maturity of practice also becomes mature). **Technology maturity** is currently expressed in terms of Technology Readiness Level (TRL). There are various TRL definitions such as defined by the US Department of Defense (Graettinger et al. 2002), National Aeronautics and Space Administration (Mankins 1995) and European Space Agency (ESA 2010). For example, Table 4.1 presents ESA's technology readiness levels. Note that the other definitions are very similar.

It is also important to better understand and assess **maturity of practice**. There is no scale yet to assess maturity of practice. When a new technology is being developed, there are fans, early adopters and people who will "buy" it no matter what; but there are also skeptical people who will find reasons to reject it. Typically, the former favor technology revolution, and the latter favor technology evolution. There is a maturity period after which it can be definitely accepted or rejected. Acceptance or rejection is related to usability, usefulness, as well as organizational and environmental integration. **Usability** is mainly related to learnability, appropriate tolerance and resistance to human errors, and ease of use (Nielsen 1993). Usefulness can be defined as "the degree to which a person believes that using a particular system would enhance his or her job performance" (Davis 1989). More objective measures of usefulness could be derived from the number of people using the new technology or system compared to the number of people who were or are still using the old technology for a given activity in a given period of time.

Is this new technology efficient in my organizational environment? Take for example email. Since it was first introduced, people have tried to learn how to use it efficiently and have run into many problems. First, it takes a while to figure out that email is both spoken language and written support for communication. This may create confusion in the way email is used, because when you write an email you may think that you speak to the person (Shea 1994), you will be sending it and you can be very casual; but when this person receives it, he or she may read it as a letter (i.e., something more formal than you expected in the first place). This may result in misunderstanding of intentions behind the email sent, and so on. Second, it is not unusual to see emails that are sent unintentionally to a list of people who should not have been recipients. This is usually due to the fast turnover of emails being read and sent; people do not take enough time to mature emails just because they are in the spoken-language mode or they commit a transmission error. Sometimes it takes several emails before reaching a goal, as in a direct conversation it would have been immediate; again this is usually due to a lack of maturity of practice.

Whenever you try to integrate a new technology into an existing organizational environment, there might be resistance from this environment and adjustments are necessary. For example, when the first glass cockpits were introduced in civil aviation, airline pilot unions were literally against this new technology. What is interesting is that a few years after the maturity period (Boy 2011, p. 432), "pilots praise(d) most of what has been done and hardly any would wish to abandon today's avionics for 'traditional' cockpit instrumentation" (Learmount 1995). One of the main problems has been "mode confusion" that typically may lead to serious incidents and even fatal accidents. Adaptation to new information technology is difficult, but we need to remember that it is a question of maturity time, both technology maturity and maturity of practice. Our society has become too impatient! In the case of the new generation of commercial aircraft cockpits, statistics show that the curve of fatal accidents tremendously decreased in comparison with conventional cockpits (Boy 2011, p. 431).

Reading the previous paragraph, you may ask, "Where is the efficiency point?" Well, even if the flight management system (FMS) and other digital technologies greatly contributed to improve commercial aviation safety, they also contributed to decrease costs such as fuel consumption and operations time. New technology is not developed without cost issues in mind; the main problem is to discover the unknown properties that will emerge from the use of that technology. I keep repeating that emergence and maturity are crucial.

Even if we do not explore and develop the concept in this book, sustainability is an important factor in human-centered design. Indeed, type of materials that are used to build systems being designed is important from various perspectives, ranging from health, to durability, recycling, and dismantling.

In many places today, most recovery techniques are reactive, and it is always better to install preventive techniques such as frequent maintenance visits and experience feedback for appropriate redesign. Working closely with domain-specific recovery professionals is certainly beneficial to improve both maintenance and redesign processes in addition to actual activities.

Comfort

The term "comfort" comes from the Latin "*confortare*", which means help, assistance, courage and aid. The "comfort" concept has evolved toward more material wellbeing factors related to the use of technology. In a study that began in 2000, focusing on comfort in a commercial aircraft passenger cabin (Dumur et al. 2004), we defined comfort both objectively (e.g., size of seats), and subjectively (e.g., pleasure and feeling of freedom). We developed four different models that are adjusted as follows:

1. the individuation model or the passenger bubble, in which the passenger is isolated from disturbances and can pursue his/her own activities;
2. the health model, where the focus is on absence of discomfort, potential health dangers and annoyance, and on physical wellbeing;
3. the community model, in which passengers belong to a public-transport group, who communicate and share common experiences;
4. the aesthetic-economical model, in which comfort is perceived as being in an interesting, advanced and beautiful environment, for a reasonable price.

Individuation, health, community, aesthetics and cost are not independent attributes. More importantly they may conflict. Design is always a question of compromises, and design teams have to weight theses attributes with respect to various principles and criteria, such as affordances, situational awareness, customization, variability and flexibility.

More generally, the concept of comfort can be modeled with respect to four disciplines:

1. **Physiology**. Comfort is defined by degree of pain and suffering, relaxation, health and physical wellbeing. Several external factors may influence physiological comfort such as temperature, pressure, noise, air quality and so on. Variability of people may differ tremendously in the acceptance of different physiological conditions. There are internal factors that influence physiological comfort such as fatigue, motion sickness, dizziness, high accelerations and so on. Biomechanics of each individual is different and will influence physiological comfort. Finally, handicapped and elderly people should have special treatment regarding physiological comfort.
2. **Psychology**. Comfort is defined by a state of quiet enjoyment, a feeling of freedom from worry, disappointment, boredom, financial difficulty, and so on. Several external factors may influence psychological comfort such as food; hygiene and security; the perception of wealth with respect to social backgrounds; aesthetics in the form of materials, colors, lights, smells and sounds; the degree of isolation

versus socialization. There are internal factors that influence psychological comfort such as ease of perception and emotions. Different comfort aspects play a role in perception.

3. **Sociology**. Comfort is defined by the conformity to groups and communities. The notion of comfort varies largely between different countries. There is a geographic and ethnic perception of comfort. Colors for example can be perceived very differently in Western and Asian counties. The notion of class in commercial aircraft creates different distinctions among the perception of comfort for passengers with respect to background and lifestyles.

4. **Technology**. Comfort is defined by technical and functional aspects of systems (e.g., visual, tactile, kinesthetic and symbolic aspects of a given life-critical system). Technological comfort is often referred to as ergonomics (Scapin and Bastien 1997; Woodson and Conover 1973), where the rational organization of various objects, devices and presentations is key (e.g., accessibility of seats, body perception of seating, acceptable noise level and operability of various instruments).

Of course, perception of comfort is influenced by education, experience, history, and living conditions. In the EURISCO[3]/Thales commercial aircraft passenger cabin comfort study (Dumur et al. 2004), using the AUTO (artifact, user, task, organizational environment) tetrahedron framework, we discovered and established the following categories that characterize comfort in a commercial aircraft passenger cabin:

- Passengers: age; gender; travel purpose (business, holidays); passenger accompaniment (alone, with family, colleagues); handicaps; experience with air travel; size plus other anthropometric and morphological features.
- Artifacts: seat: including armrest, headrest, footrest, controls; table; seat belt; light; IFE (In Flight Entertainment) plus control panel, including music, video, games, weather report; flight information; laptop connection; internet; telephone; air-conditioning; luggage compartment; journals and books; food and drink tray; crew alert; smart card reader.
- Activities: sleeping; reading; working (with a laptop or other personal electronic tool; writing, with paper and pencil, reading document); being entertained and playing (music, video, games); eating; communicate with others passengers, with crew and with exterior; move around the cabin; act on seat, environment and IFE parameters; follow security demands: fasten seat belts, listening to security announcements; shopping (tax free, services).
- Environment: crew; other passengers; decoration; flight duration; point of departure and destination; connections; cabin layout; turbulence; physical conditions within the cabin (temperature, light, noise, air quality).

---

[3] European Institute of Cognitive Sciences and Engineering.

Comparing Life-Critical System Domains

Even if the following LCS domains are in different stages of evolution, let us analyze and compare them.[4] The categories are aviation, medicine, nuclear power plants, the automobile and human spaceflight.

Common properties of these domains are their increasing complexity, the cumulative layers of software, the difficult problem of functions allocation between humans and systems that currently lead to **human-system integration** issues, national and international regulations that lead to more procedures and rules, risk taking and management, resilience and socio-cognitive stability, emergence of new skill sets, in addition to rules and knowledge from increasingly automated systems. Consequently, automation requires proper cognitive function analysis that includes authority and responsibility analyses. These analyses should now be carried out in areas where systems are not necessarily easy to automate.

The issue of socio-cognitive stability requires HCD teams to propose systems that can recover from failures and return to safe and successful operation. Various kinds of defenses are put into place to attempt to protect LCSs against failures but we cannot simply rely upon them. This may lead to a false sense of safety (i.e., defenses are not only technological); they also are part of the constant involvement and engagement of human operators and technical managers. In particular, we need to be aware of the trend leading to failures or improper operation being undetected for longer periods of time.

All these domains are typically developed in large industries with similar goals. Engineers and companies, which design these LCS systems, require an extra level of attention to ensure safety of end users as well as all actors dealing with such systems.

Commercial Aviation

In our modern socio-technical world, the aviation industry has the longest history of LCS design with the human in mind. For most tasks today, commercial aviation belongs to the class of routine work. Operational procedures, checklists and do-lists support such work. Operational procedures can be paper-based or electronic. Airlines pilots are expected to read checklists, which have been developed to help them stay "ahead" of the aircraft. Procedures enable the aircrew to anticipate actions and avoid forgetting to do what has to be done (Boy and DeBrito 2000). More generally, such procedures are usually pre-programmed sequences of elementary tasks that human operators should execute sequentially to accomplish specific goals in a specific environment; for that matter, they tend to "automate" human work.

---

[4] My graduate students did a great job in providing excellent term paper content that I used in this section of the book. Thanks to Sharon Chinoy, Carol Craig, Rhonda Lyons, Don Platt, Kara Schmitt and Lucas Stephane.

Fig. 4.3 Example of a procedure that enables a pilot to identify if the manual trim is available or not; actions are different in each situation. (Reprinted from Boy and DeBrito 2000)

STABILIZER JAM
- MAN PITCH TRIM.. CHECK
■ **IF MAN TRIM AVAIL :**
- TRIM FOR NEUTRAL ELEV
● **if normal law lost :**
F/CTL ALTN LAW (PROT LOST)_ See previous page
■ **IF MAN TRIM NOT AVAIL and normal law active :**
- LDG CONF 3.. SELECT
- GPWS LDG FLAP 3 .... ON
- etc.

Procedures are presented in the form of simple lists of actions or more sophisticated algorithms that users should perform (Fig. 4.3).

Operational procedures are good to keep at an acceptable level of vigilance, and ensure that all required actions are performed in the right order and at the right time. We found out that some procedural items may not be executed when pilots did not understand their rationale in context (De Brito et al. 1998). Indeed, even if aviation procedures are incrementally refined from constant experience feedback, they may not be always valid in an unusual or unanticipated context. A pilot's airmanship is still mandatory to handle the unexpected (Pinet 2011). Note that both aircrew and air traffic controllers must commonly have a high level of training and experience. Aviation experience feedback is very well organized and used. A safety culture is in place. In addition, there are families of aircraft; this "standardization" tends to improve operations efficiency and safety. Phases of flight are also very well categorized, and context is well mastered (e.g., take-off and landing are the critical phases).

Practices are the actual forms of crew interaction with systems. These practices need to be fully accounted for in the process of procedures development by designers and operators of new generation aircraft. This implies that the policies, by which manufacturers define how automation is implemented in design and by which operators specify the way to perform in the cockpit, are clearly adapted to the needs of pilots. Policies, however, derive from an overall philosophy, by which management defines the overall way to include automation in the design process and how to conduct the airline, in particular by defining flight operation criteria. The 4Ps model (i.e., philosophy, policies, procedures and practices) was developed during the early nineties (Degani and Wiener 1994). These 4Ps are handled at different technical and administrative levels among aircraft manufacturers, airlines and national organizations such as the Federal Aviation Administration (FAA) in the United States, and the International Civil Aviation Organization (ICAO) at the world level.

Today, aircraft cockpits have evolved from levers, dials and electro-mechanical instruments to digital technology, enabling more capabilities in terms of safety, efficiency and comfort. A commercial aircraft cockpit looks more like a virtual office since the desktop metaphors are all there, transplanted. Is it good or not? This is a question that deserves more investigations. This new generation is called "interactive cockpits !" Were cockpits not interactive before? Well, there are two types of interaction in this case: continuous/analog interaction with flight mechanics, and discrete/digital interaction with information technology. The term "interactive cockpit" refers to the latter distinction.

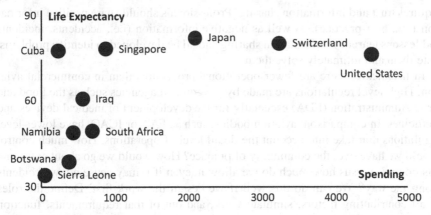

Fig. 4.4 Life expectancy (years) vs. per capita health care spending (international US dollars). (Adapted from UC Atlas 2012)

Glass cockpits were designed in the early eighties; they are now fully adopted in commercial aviation (i.e., the maturity period is over). However, the new generation of interactive cockpits will very likely change practices again, and a new maturity period has to be expected (Boy 2011, p. 432).

Medicine

The University of California Atlas of Inequality,[5] led by Dr. Ben Crow, recently provided a map of life expectancy versus health care spending (Fig. 4.4). You can see that the United States of America spends a large amount of money in health care for a life expectancy that is even lower than other high performance level countries such as Japan. In fact, this curve does not show intra-country inequality (e.g., ethnicity, people with lower incomes and less education tended to die younger). In occidental countries, current healthcare organizations (i.e., technologically advanced systems) show many human errors committed by intelligent, motivated and hard-working people. Why?

Medicine and more specifically the healthcare industry have a very long history that needs to be traced and shared among experts. As in aviation, human reliability is at stake as well as human involvement and engagement. System and human reliability is mostly improved from experience feedback, but experience feedback cannot be done correctly if people fear legal issues when they attempt to share what could be interpreted as mistakes. Consequently, the issue today is the domination of very procedural precautions triggered by lawsuits. Conversely, installing a safety culture

---

[5] The Atlas integrates data, maps, and graphs to create an interactive website for accessing and analyzing information addressing global change and inequality.

requires trust and information sharing. Professionals should share positive information (i.e., best practice), as well as negative information (i.e., accidents, incidents and lessons learned). Information sharing should be developed to identify problems, state them and ultimately solve them.

In healthcare, there are fewer operational procedures than in commercial aviation. High-level regulations are made by government agencies such as the Food and Drug Administration (FDA) especially for the development of medical devices and medicines. In comparison, aviation bodies, such as FAA or ICAO, have lower-level regulations that take into account the detail level of operations. How much control should we have over the community of practice? How would we go about enforcing this control, versus how much do we allow it, even if it may cause a few accidents along the way? We can do this well if we design the work first. Define the roles of all contributing factors, simulate work, and out of that requirements, function allocation, conflicts, and so on.

Indeed, very little is done regarding experience feedback (i.e., with respect to formal incident investigations and sharing of safety information within the medical domain). Accidents are often kept hidden as much as possible to limit possible litigation. Medical technology usability is not systematically tested, as it would be in aeronautics. Compensation is found in the level of technical qualification of personnel. Unfortunately like in many industrial sectors, finance-driven management is often the main issue with management personnel who do not have clinical experience to make decisions.

Automation is less developed in medicine than in other life-critical domains. In contrast, there are too many alarms that physicians have to monitor. The main issue is that alarms are often ignored simply because there are too many or the alarms get turned off as in many cases (Wears 2011). Personnel turnover is also high so documentation is not done properly in terms of state of equipment or how equipment works. Confusion can also be generated from medicine bottle labels that are wrongly assigned, or dosage and concentrations of medicine that are not clearly marked. In a busy hospital, it is easy to grab the wrong medicine.

Health care professionals have recognized that they need help in creating a safer industry and have talked about collaborating with outside experts from other domains but to date little real progress has been made. Human-centered designers are called in after the fact to try and fix problems rather than be consulted from the beginning of a new system design.

Automobile

Unlike aviation and medicine, automobile's operators are not necessarily experts. Driving a car is a general public act. In addition, an automobile does not require paper-based procedures and checklists to follow, even if there are road rules that drivers are required to follow. These road rules are established at both national and international levels by transportation safety authorities. The way people obtain their driver licenses varies from country to country. For example, you can obtain your driver license when you are 16 in the United States of America, and 18 in France. Examination requirements also vary.

Driving requires constant attention and anticipation. These kinds of cognitive functions are acquired with experience. There are various sorts of situation awareness problems in driving, such as speed control, trajectory control in various road conditions, and multi-tasking. These problems may be increased by distraction, fatigue and absorption of alcohol or drugs, for example. A good example is the integration of the telephone in a car. In most countries, the use of cell phones is prohibited while driving. Why? It involves multi-tasking that requires skills regular drivers do not usually have. Multi-tasking could be performed at different behavioral levels such as the perception-action level where it is difficult, and most of the time impossible, to dial and change gears at the same time, or at least coordinate these actions in a large set of situations. Multi-tasking is even worse at the cognitive level where a phone conversation may concentrate attention and distract navigation tasks for example. In fact, talking on the phone and driving at the same time are disjoint tasks, which require specific training. Phoning while driving induces a new concept, "the communicating car" that requires further human and machine adaptations (Morel 2008).

As in other domains, the automobile has become more automated (Bellet et al. 2011). If fly-by-wire was designed and developed during the eighties, we are in the middle of drive-by-wire development today. Information technology is used for various intra-vehicle purposes such as speed and trajectory control, situation awareness using proximity sensors and cameras, and car maintenance management. For inter-vehicle purposes, information technology can be used for navigation and collision avoidance for example. This should not be limited to vehicle equipment; collaborative road infrastructure is certainly a major endeavor in the near future.

Human-centered design of future automobiles should then be focused on a more integrated approach than what we can observe today. Indeed, many assistance systems have been developed such as GPS, speed control system, collision avoidance system, and telephone hands-free-kit; they can be useful individually, but they also can be very disruptive when they are used all together because drivers are involved with additional multi-tasking activities. For that matter, integration is key; it involves an AUTOS pyramid approach (explained in Chap. 7 on Modeling and Simulation).

Nuclear Power Plants

As stated earlier (Fig. 4.2), the nuclear industry is certainly an ultra-safe LCS industry with only three major accidents in more than 40 years of operations around the world. Nevertheless, nuclear systems have been quoted to be "inherently complex, tightly coupled systems where, in rare emergency situations, cascading interactions will unfold very rapidly in such a way that human operators will be unable to predict and master them" (Perrow 1999).

Fig. 4.5 Automation of
people and machines

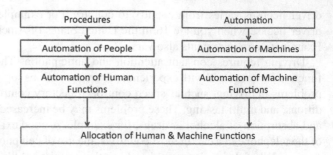

A nuclear power plant (NPP) needs to be licensed/certified by Safety Regulatory Bodies. Licensing requirements vary from country to country, even within the
international framework of the International Atomic Energy Agency (IAEA). A major difference can be mentioned where in France the *Autorité de Sûreté Nucléaire*
(ASN) requires that NPP operators provide demonstration (result obligation), and
in the United States the Nuclear Regulatory Commission (NRC) requires that NPP
operators respect defined rules (means obligation). In the US, the licensing process is
very well defined under many Codes of Federal Regulations (10CFR). For example,
human factors and ergonomics rules are defined in 10CFR55.

Typically, a NPP operator must have 3 years of power plant experience prior to
beginning training, and 1 year at least in the NPP where the operator plans to be
licensed. Training simulators are used like in aeronautics.

Even if commercial aviation is very procedural, the nuclear energy sector is certainly the champion in the production of regulations, rules and procedures. This is
due to very well established experience feedback processes. Today, the number of
operational procedures in a nuclear power plant is huge and deserves more integration
for safer, more efficient and comfortable usage.

NPPs are very automated. Consequently, a NPP human operator's job is mostly focused on monitoring, except during starts, incidents and accidents, and shut downs.
Automation has become an accumulation of software that assumes the burden of
tasks previously executed by people. We do not need to be surprised by this evolution since, when people use procedures, it is **automation of people** (Fig. 4.5).
What is the difference? People are more flexible than machines when they follow procedures; they are context-sensitive. Both automation and procedures are
good for routine activities in general. When both automation and procedures become mature (i.e., technology and practice maturity), human operators use them
as safeguards (i.e., cognitive support for cognitive stability assurance). The industry application of contextualization provided by electronic procedures is very
promising (Ramu et al. 2004; Ramu 2008). Contextualized electronic procedures
can be considered as machine agents that support human interaction with complex
systems.

Why do we talk about "automation of people"? In psychology, Shiffrin and
Schneider (1977) already introduced the distinction between automatic and controlled human behaviors. They argued that human automatic processing do not

require much attention; instead, human controlled processing, such as learning, require full attention from the subject. Automatic processing is based on skills. We can compare Shiffrin and Schneider's automatic processing to Rasmussen's skill-based behavior (Rasmussen 1986). Controlled processing may become automatic through learning, training and experience; we can talk about skill acquisition, which is a kind of automation. In addition to skills, Rasmussen's behavioral model brings another distinction (i.e., rule-based and knowledge-based behaviors). As we talk about rule-based behavior, we also can talk about procedural behavior. For that matter, this behavioral level can be modeled by procedures that can be learned by people or explicitly used as external support. Paper or electronic procedures are nothing else more than external memories for human operators. Whether they are internal or external, they automate human behavior at the second behavioral level of Rasmussen's model. The main difference with skills is that the procedural level is conscious. What remains of course, is the knowledge-based level used when both skills and procedures fail. Consequently, in complex systems, such as nuclear power plants, human operators should be highly equipped with such a knowledge-base level in addition to others.

Human Spaceflights

As in the other industries, safety is the number one goal in human spaceflight. Human spaceflight has the smallest statistical set of operations since very few vehicles have been built and flown so far and less than 1,000 humans have actually flown in space since Yuri Gagarine in 1959. Clearly risk is still very high since the environment is completely unforgiving of any errors or failures.

There were 135 space shuttle missions in 30 years, and two fatal accidents, Challenger in 1986 and Columbia in 2003. It is interesting to say that at any point, any employee can stop a launch, assuming there is a safety reason to do so. From there an investigation is performed to verify all items noted are addressed and completed prior to a launch.

There are many procedures and checklists in place for human spaceflight. Human spaceflight is marked by having each moment of flight be rehearsed and trained for by a factor of as much as 10 (i.e., for each hour of operation there are 10 h of training).

There is more automation in human spaceflight now than in the Apollo days. As we move farther from Earth this trend will continue. Crews so far have relied upon a constant communication string with mission control on Earth but distance will make this impossible at Mars for instance. Decision support and knowledge enhancement tools such as the Virtual Camera will be critical to assist crews in completing these remote missions. The virtual camera concept will be described in Chap. 8.

So far there are fewer regulatory elements for human spaceflight although this may change as more commercial entities become involved in the domain. There are new FAA and medical requirements currently being defined for commercial human spaceflight including space tourism.

Table 4.2 Life-critical system domain comparison

Domain	Space	Driving	Medicine	Aviation	Nuclear
Procedures	Medium	Low	Low	High	High
Regulation	Low	Medium	High	High	High
Safety culture	Medium	Low	Low	High	High
Training	High	Low	Medium	High	High
Automation	Medium	Low	Low	High	High
Safety level (hr^{-1})	10^{-2}	10^{-4}	10^{-4}	10^{-6}	10^{-6}

In terms of learning from accidents, when comparing the shuttle Challenger and Columbia accidents, many of the same errors were made once again although the two accidents were separated by 17 years. New systems being developed for human spaceflight are concentrating on failure recovery rather than the idea that the system is reliable and an accident will not happen. The "faint" signals that an accident may be impending must be heeded. NASA is doing this more now by looking more closely at engineering data and not simply thinking that "we got away with it before so it will be OK to fly again."

Domain Summary

All of the domains described above have fairly detailed processes in place for design and development of the respective systems. These include certified design steps and sequences. As with any certification process, a key is to not simply believe the end result will be mature and safe because the process is followed. However, design tools and processes can aid in proper documentation of design as well as aid in the iterative process of prototyping, simulation and testing, which can culminate with a mature design.

Table 4.2 summarizes the various life-critical domains discussed. It can be seen that certain characteristics are common and certain ones differentiate various life critical domains. Safety level is the number of accidents per hour of operation. It can be seen that domains ranked high in all categories are the two "ultra-safe" domains described above. Human spaceflight is difficult to rate in terms of accidents per hour of operation so the safety level is really more reflective of accidents per flights for the space shuttle. Training may not offset other issues inherent in system operation. Spaceflight has high training levels but yet is still a proven "unsafe" activity because it is extremely complex.

As in any system designed, developed and operated by humans, organizational issues have to be considered in life critical systems. Many of the current failures and shortcomings that have been seen in all of the industries under study can be traced back to organizational issues such as placing financial decisions above engineering and safety concerns, asking personnel to do too much for too long a period of time, and trying to put band aids on problems after the fact. Any organization that fails to understand the life-critical aspects of what they are doing will experience an accident at some point of time.

Recommendations

Designing life-critical systems requires principles and rules that address the following questions. How can machines assist people in their jobs and activities? How can we take advantage of the unique abilities of the human to recognize and solve problems? First, safety is a deeper concept that cannot be simply taken into account at the user interface level once a system is developed. Second, people are not only problems because they commit errors; they can be solutions because they are unique problem solvers for unforeseen situations. Third, experience is key and must be enhanced at the individual level by promoting long-term involvement and engagement, as well as at the collective level by promoting structured and integrated experience feedback processes.

There is no such thing as a zero-risk system or practice (Boy and Brachet 2008). Therefore, we must educate the public about risk. Risk is part of life and we must deal with it and embrace it. We need to better understand risk, not just by probabilities, but also by a deeper understanding of its nature, and ultimately better, more stable system designs. People should be allowed to make mistakes and report them to make the community aware of various dangers. Error tolerance requires concrete possibilities of recovery, and in many cases, competent and knowledgeable people should be given a chance to recover from errors instead of assisting them systematically.

Designing a system requires a deep understanding of the work it will support. The following questions need to be addressed at design time, as well as during the whole life cycle of the system. What work needs to be completed? How can we lower workload and not impose new burdens on the operator, induced by the system being designed? How can we properly allocate authority between humans and machines? Who is in charge and who is legally responsible? We should not automate and just leave the human as the backup device. How can the human take over a system that he or she has not been actively involved in and do it at the worst possible time, when a failure has occurred? Moreover, we need to anticipate this possibility of failure and provide human operators the means for appropriate recovery.

In human-centered design, both creativity and experience are equally important. It cannot be conceived otherwise than in a team involving various kinds of expertise. Scenario-based design, iterative evaluations and human-in-the-loop simulations are crucial in the human-centered design of life-critical systems. Most importantly, management decisions should not simply be governed by financial considerations. Do not forget that designing new systems is always a matter of compromises, and when life is at stake, appropriate principles and criteria must be used.

Safety Models

In order to understand a life-critical system, we must understand the risk it involves and how to mitigate this risk. Risk can be divided into categories of intentional and circumstantial risks. Intentional risk can be controlled, and is familiar and calculated. Circumstantial risk is often unforeseen and a sequence of uncontrolled events, such as weather. However risk comes, when a dangerous situation occurs, we must plan for it. Although risk usually has negative connotations, it can also be good. It is important to understand not only the negative outcomes, but also the positive ones.

How can risk research be developed? Risk research cannot be carried out without deep immersion in life-critical systems themselves, and development of descriptive models. Modeling within a system allows us realize all of the little things we would not have understood otherwise. Technical models such as AltaRica, an example of mathematical model-based reliability engineering can help us to understand the way a system interacts with other components (Rauzy 2008). Modeling and simulation can certainly break a system down into understanding, help to assess risk and reliability of a system, thus allowing us to see errors that may occur.

The AltaRica model is general enough that it can span multiple domains, but is targeted enough that engineers can implement specifics of a nuclear power plant, chemical plants, aircraft, and defense systems, for example. Syntax based programming language models reliability and safety of specific components within the systems, that can then assess risk and answer the three basic questions of any risk assessment as stated above. However, we face issues of modeling people. How exactly can you model a user that has emotions and thoughts? A model-based designed system can be useful, though depending on how it is executed it can easily have limitations. Often times, when we look at modeling the system, it allows us to better understand the technical aspects and interactions, but not necessarily the cognitive functions, or how the users fit in. Though an engine can be successfully modeled, the human element cannot be fit into a box to be predicted, because they are unpredictable based on infinite number of variables and situations. Two people will not react the same to the same situation. Even one person may react differently in the same situation given a chance to do it again. Descriptive models improve understanding of human capabilities, but it is almost impossible to use human predictive models except in very simple situations and for very simple behaviors.

In addition to modeling, we may also utilize probabilities and statistics to understand a system. We can look at past data, accident data and incident data, in order to determine what went wrong and how to improve it. The difference from this and modeling is the use of real statistics that modeling can only predict. In this chapter, the fatality rates within the aviation, medicine, nuclear and automobile industries were compared. By looking at real world historical data, we can begin to understand the emergent functions within a system that may not have been apparent during the design phase. I claim that accidents usually occur when we do not understand the emerging cognitive functions, or the how the automation works when either technological maturity, maturity of practice or both are not reached. Again, maturity is a very important concept that needs to be mastered and properly assessed.

Public perception is very important within any industry with regard to safety. Activist groups do have an impact on industry regulations and growth. The differences between actual risk and perceived risk are astounding. People who are afraid of flying should know that automobile accidents kill more people than aviation. There is a phenomenon that people as a whole do not seem to mind losing one life or two, but if there is an airline accident and 200 people perish, it makes headline media.

Dealing with the Unexpected

Life-critical systems are strongly related to risk, whether it is risk in making them or risk in using them. The notion of hazard is often suggested to characterize these systems as a property of a substance or situation with the potential for creating damage. Risk itself is typically defined as the likelihood of a specific effect within a specified period. Risk is a complex function of probability, consequences and vulnerability.

Several methods have been developed and used to assess safety in complex dynamic systems. Among them, Probabilistic Risk Assessment (PRA),[6] is widely used in safety-critical industries. PRA is based on two entities: the probability of an accident occurring (P_A); and the expected loss in case of accident (L_A). It is usually implemented in the form of a product (i.e., risk $= P_A * L_A$). Think about an accident such as in Fukushima Daiichi in 2011. P_A was extremely small, and L_A was huge. What do you think would have happen if we did use the above risk formula to anticipate the risk in Fukushima Daiichi? We learned in mathematics that zero multiplied by infinity is undetermined, in other words the risk in Fukushima Daiichi was undetermined. Therefore, approaches such as PRA work well on the Gauss bell curve, but not on the small residual parts of this curve. The problem is that we operate precisely on these residual parts.

A tremendous amount of work has been done in the field of system safety, reliability, availability and dependability (Johnson and Malek 1988; Laprie 1992, 1994; Prasad et al. 1996; Nilsen and Aven 2003). There are methods that were developed to assess human reliability as an extension of probabilistic methods addressing system reliability such as the Technique for Human Error Rate Prediction (THERP) (Swain and Guttman 1983) or Standardized Plant Analysis Risk (SPAR) (Gertman et al. 2005). The main issue is that probability barely works for predicting human errors and more generally human behavior. Again, these kinds of approaches are not appropriate for dealing with Black Swans (Taleb 2007). The Black Swan theory was developed by Nassim Nicholas Taleb to address unexpected events. These events are impossible to predict and therefore we should not attempt to predict them. In contrast, we should develop robustness against negative events and exploit positive ones. Unexpected situations and events require training and expertise to handle them. Underlying knowledge is in the understanding of complexity theory, as well as familiarity to cope with the unexpected, dealing with emotion in particular.

Recently, I chaired the session on "pilots' reaction to the unexpected" during a conference in Paris organized by the Air and Space Academy and very interesting conclusions emerged. We should develop an appropriate mix of operational knowledge and experimental information, associated with social, cognitive and technical knowledge, leading to a better grasp of complexity of the life-critical system that is the current airspace of commercial aviation. Complexity is increasing because the

---

[6] The following URL provides useful links on PRA: http://nuclearsafety.info/probabilistic-risk-assessment/.

number of components increases, and more importantly the number of interconnections among them increases also. These components can be aircraft, and software agents of any kind. For these reasons, authority sharing and function allocation have become crucial issues. However, one thing that keeps being critical is competence of the various actors. This is related to both initial and recurrent training.

Life-Critical System Stability

Hollnagel, Woods and Leveson introduced the term **Resilience Engineering** in 2006 to denote ways for enhancing the ability of organizations to create robust and flexible processes, to control and manage risks, and to use resources proactively when things are going wrong or when there is too much pressure. This approach departs from the traditional limited human reliability approaches. Resilience Engineering defines itself as the necessary adaptation of socio-technical systems to cope with internal or external risks.

The concept of resilience is interesting since it implicitly put human-centered design of technology and organizations first. It moves from the negative view of human-machine reliability to the positive view of adaptation. However, I prefer the concept of **stability** that is more appropriate for multi-agent systems. Indeed, stability of a multi-agent system results in constant co-adaptation that results in emerging patterns or attractors. We need to better understand these attractors. Consequently, we need to address the difficult issue of LCS complexity. As we will emphasize in Chap. 5 on the making of complex systems, complexity can be situational and organizational. Situational complexity deals with the unexpected and various reactive behaviors that result from it. Organizational complexity deals with connections and interactions among various agents in the socio-technical system being studied. In both cases, any interaction may cause disturbances that amplify. In complexity science, and chaos theory in particular, it is stated that a small variation on an input may cause very unpredicted results. The problem is not so much to study individual trajectories that are sensitive to initial conditions, but focus on attractors that we usually consider as emergent properties.

As already said, the concept of stability comes from physics where passive stability and active stability are clearly defined (Boy 2001). A passively stable system returns to its stable state when disturbed (e.g., the pendulum). An actively stable system diverges from its stable state when disturbed (e.g., the inverted pendulum). When you interact with a passively stable LCS, you may commit errors that do not create major changes or disturbances. Conversely, when you interact with an actively stable LCS, any error will have to be compensated in order to maintain a stable state. The type of stability of a system is something that you need to know. When you design a new system, even if you always try to anticipate its use, you do not know how people will use it until they actually do. This is why modeling and simulation are so important in searching for emerging properties before product delivery. Once you know them, you have some information to re-design the LCS to cope with disturbances either passively or actively.

Fig. 4.6 Training and
operations situations

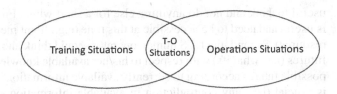

Major Operational Processes

Three major operational processes will be presented and discussed: training,
experience feedback and crisis management.

Training

In human-centered design, we try to adapt technology and organizations to people.
However, people also need to adapt to evolving technology and organizations. Consequently, training is always required, and in very specialized jobs, selection needs
to be done. In an LCS such as aviation, pilots are trained to always comply with predefined procedures and demonstrate that they are able to correctly execute standard
exercises. Nothing must be left at random. Everything is codified. However, we have
no guaranty that pilots will know how to deal with an expected situation. We use
simulators to train pilots for situations that they may never face during their whole
professional life. The main challenge is to maximize the T-O situations set, that is
the set of relevant operational situations we can anticipate and use at training time
(Fig. 4.6).

It is often the case that the T-O situations set is incrementally improved and refined
from the delivery of a new aircraft to operations times. In other words, the more we
use a system, the more the T-O situations set is maximized. The question is then:
how do we train pilots to handle unexpected situations?

If we use the model of Rasmussen, unexpected situations deal with the knowledge-
based level. The skill-based level is incrementally constructed from good training in
well-known situations in order to equip pilots with appropriate reactions to disturbances. The rule-based level is typically supported by operational procedures that
are both learned by pilots and used as external memories to react to specific situations. At the knowledge-based level, pilots have to identify an actual situation, make
a decision and plan an organized set of actions. Identifying a situation that is not
expected either because we do not think about it but we have seen before is different
from identifying a situation that is not known at all. In the former case, it would be
useful to have an automated system that provides the pilot with appropriate situation
awareness. Let us provide a few heuristics.

First, flying is not a natural activity for human beings. Consequently, intuition
is not a good friend. In particular, when an unknown situation occurs, stress and
emotion are very likely to increase. In this case, before doing anything, it is always

useful to think and not do anything else for a short while. First principle knowledge is useful and need to be accessible at this time (e.g., flight mechanics is a mandatory resource). Second, the LCS human operator should think ahead, investigate possible futures (i.e., what ifs) with respect to his/her available knowledge, and choose one of possible futures according to currently available information. Third, critical thinking is crucial (i.e., any contradiction in available information should be immediately checked and appropriate reaction implemented).

Today and likely in the future, access to knowledge is often much easier and faster than it used to be. Both databases and simulation capabilities are far more powerful than ever before. Therefore, training should be focused on using these new capabilities. Of course, this does not remove keeping high competence both onboard and on the ground, and training to work cooperatively (i.e., computer-supported cooperative work training). We can generalize this position to other life-critical systems where the various actors must be identified and properly trained both individually and collectively.

Experience Feedback

It follows that experience feedback is very important. However, this process should be done with integration in mind. There are several systems such as the Aviation Safety Reporting System[7] that enable "the identification of deficiencies and discrepancies in the National Aviation System (NAS) of the United States so that these can be remedied by appropriate authorities." It also enables "supporting policy formulation and planning for, and improvements to, the NAS." Finally it enables "strengthening the foundation of aviation human factors safety research. This is particularly important since it is generally conceded that over two-thirds of all aviation accidents and incidents have their roots in human performance errors."

However, when such systems are extensively used, they lead to production of a large set of regulations and consequently operational procedures. As observed in the nuclear domain, **drowning into the excess of procedures and rules** is also a problem that arises from the implementation of very systematic experience feedback mechanisms. Such mechanisms usually record **failure-based experience** (i.e., what did not work well such as incidents and accidents), instead of success-based experience (i.e., what worked well such as well-planned and executed tasks), and more importantly some well-recovered abnormal and emergency situations. In order to avoid the occurrence of such failure-based experience in the future, people tend to develop new sets of constraints in the form of regulations, procedures and rules. This is the case in safety-critical industries such as nuclear energy and aeronautics. We observe that the experience feedback homeostatic loop tends to become positive as presented in Fig. 4.7 (i.e., progressive accumulation of operations procedures and

---

[7] http://asrs.arc.nasa.gov.

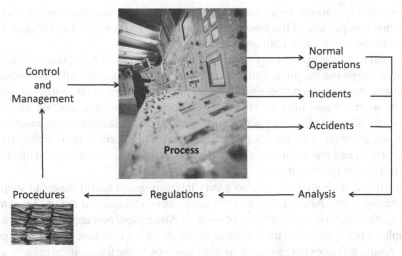

Control
and
Management

Process

Normal
Operations

Incidents

Accidents

Procedures ← Regulations ← Analysis ←

Fig. 4.7 The homeostatic positive loop of the accumulation of procedures. (Stock image courtesy of Digital Vision/Thinkstock)

rules) instead of negative (i.e., redesign of the whole system with improved function allocation among the various human and machine agents).

Experience feedback should then be coupled to cognitive function analysis (CFA) and the redesign of technology, organizations and people jobs (TOP). This is the price to pay in order to achieve socio-cognitively stable LCSs.

Crisis Management

Even if we always try to avoid accidents, we also need to plan for crisis management in case of serious situations, such as the Apollo 13 successful accident,[8] and the way all involved actors cooperated to solve the problem and bring the space crew safely back to Earth.[9] The crisis during the mission has been reported in a number of books and popular movies. Landing on the Moon was aborted after an oxygen tank exploded. The crew lost cabin heat, had limited power, shortage of water and a dysfunctional dioxide removal system. Success of the recovery from this very unexpected situation was mainly due to cooperative work and competence of the crew both on board and on the ground. In particular, people on the ground found a way to power up the Command Module with its limited available power. They actually simulated processes on the ground and sent procedures to the Apollo 13 crew. This very interesting experience shows that simulation can also help in real time during

---

[8] http://nssdc.gsfc.nasa.gov/planetary/lunar/ap13acc.html.

[9] This example is crucial since there are very few reports on successful accident (positive experience). Indeed, experience feedback usually provides accidents that led to casualties (negative experience).

safety-critical operations. From a training perspective, success came from a powerful leadership, competence of the whole group of support staff involved and cooperative training of the whole crew both on board and on the ground.

We need to take into account that people have to process a large quantity of information, manage the magnitude of the situation and more importantly the surprise. In addition, information processing is in real time in an off-standard situation where data are uncertain and time is critical. Several kinds of management are involved such as management of various actors (intersubjectivity), management of communication, management of decision impact, management of information credibility (trust maintenance), and management of personnel (workload and health in particular) and collective image (public impact).

Modeling and simulation are also a very interesting and useful means to support crisis management. Agent-based models are great candidates and are used to simulate possible futures (i.e., simulation of what if). Since these new capabilities are now available, it is important to train experts to use them for real-time operational purposes. Again, this does not remove educated common sense for decision makers who should not believe simulation results blindly. We must never forget that technology is a support mechanism and does not provide a responsibility transfer.

Conclusion

Various life-critical domains were compared in this chapter. Those ranging from "safe" to "ultra-safe" have many procedures and high levels of regulation. This is not to say that just by applying procedures and regulation you will automatically have a safe system, but procedures are most of the time good barriers against routine errors. Sometimes procedures can be overburdening and limit the ability to solve problems. Life-critical systems should enable involved people to solve problems in best possible conditions when necessary. Remember that the human element is not a black box that can be put in chart in the middle of other boxes representing systems; the human element is very complex and deserves full attention.

Life-critical systems are all evolving to higher levels of automation. We cannot assume that higher automation will automatically mean safer systems. Automation has to be done in an intelligent way. In particular, instead of accumulating layers of software on top of each other, it is crucial to focus on integration anytime a new design is at stake.

In addition, there are other types of LCSs that deserve more attention. For example, the cell phone has become a life-critical system because it is part of our everyday life for three main reasons: for safety reasons because we can call or be called for assistance in case of danger or emergency any time anywhere; for efficiency reasons, the cell phone enables rapid, effective and efficient problem solving, and facilitates collaborative work; and finally for comfort reasons because the cell phone provides comfortable direct link with our extended environment—we know that we are no longer isolated—thus providing "peripheral socio-cognitive stability". Consequently, the cell phone has generic emergent properties that impact safety, efficiency and comfort.

References

Amalberti, R., Auroy, Y., Berwick, D., & Barach, P. (2005). Five system barriers to achieving ultrasafe health care. *Annals of Internal Medicine, 142*(9), 756-764.

Bellet, T. (2011). *Road safety: From car driver modeling to advanced copilot.* Lecture given at Florida Institute of Technology, Melbourne, March 4.

Bellet, T., Hoc, J. M., Boverie, S., & Boy G. A. (2011). From human-machine interaction to cooperation: Towards the integrated copilot. In C. Kolski (Ed.), *Human-computer interactions in transport* (pp. 129–155). London: ISTE Ltd and Wiley.

Boose, J. H., & Gaines, B. R., (Eds.) (1990). *The foundations of knowledge acquisition, knowledge based systems* (Vol. 4). San Diego: Academic.

Boose, J. H., & Gaines, B. R. (Eds.). (1988). *Knowledge acquisition tools for expert systems, knowledge based systems* (Vol. 2). San Diego: Academic.

Boy, G. A., & de Brito, G. (2000). *Towards a categorization of factors related to procedure following and situation awareness.* Proceedings of HCI-Aero 2000. Toulouse: Cepadues.

Boy, G. A. (2001). *When safety is a matter of redundant information.* Proceedings of HCI International, New Orleans, August.

Boy, G. A. (2011). Cognitive function analysis in the design of human and machine multi-agent systems. In G. A. Boy (Ed.), *Handbook of human-machine interaction: A human-centered design approach.* Aldershot: Ashgate.

Boy, G. A., & Brachet, G. (2008). *Risk taking. Dossier of the Air and Space Academy.* Toulouse: ASA.

Bradshaw, J. M., Ford, K. M., Adams-Webber, J. R., & Boose, J. H. (1993). Beyond the repertory grid: New approaches to constructivist knowledge acquisition tool development. *International Journal of Intelligent Systems, 8,* 287–333.

Davis, F. D. (1989). Perceived usefulness, perceived ease of use, and user acceptance of information technology. *MIS Quarterly, 13*(3), 319–340.

De Brito, G., Pinet, J., & Boy, G. (1998). Etude SFACT sur l'utilisation de la documentation opérationnelle dans les cockpits de nouvelle generation (new generation cockpit operational documentation study). Final Report SFACT No. 94002, Paris: DGAC.

Degani, A., & Wiener, E. L. (1994). *On the design of flight-deck procedures.* NASA Contractor Report 177642, Moffett Field: NASA-Ames Research Center.

Der Kiureghian, A., Ditlevsen, O. D., & Song, J. (2005). Availability, reliability and downtime of systems with repairable components. *Journal of Reliability Engineering & System Safety, 92,* 231–242. March 2005.

Dumur, E., Barnard, Y., & Boy, G. A. (2004). Designing for comfort. In D. de Waard, K. A. Brookhuis & C. M. Weikert (Eds.), *Human factors in design* (pp. 111–127). Maastricht: Shaker.

ESA. (2010). http://sci.esa.int/science-e/www/object/index.cfm?fobjectid = 37710. Accessed 22 Jan 2012.

Fitts, P. M. (1951). *Human engineering for an effective air navigation and traffic control system.* Washington, DC: National Research Council.

Gertman, D., Blackman, H., Marble, J., Byers, J., & Smith, C. (2005). *The SPAR-H human reliability analysis method.* NUREG/CR-6883. Idaho National Laboratory, prepared for U. S. Nuclear Regulatory Commission.

Graettinger, C. P., Garcia, S., Siviy, J., Schenk, R. J., & Van Syckle, P. J. (2002). Using the technology readiness levels scale to support technology management in the DoD's ATD/STO Environments: A findings and recommendations. Report Conducted for Army CECOM. Special Report, CMU/SEI-2002-SR-027, September, CMU Software Engineering Institute.

Hansman, J. (2011). *Automation issues in advanced cockpits.* Lecture given at Florida Institute of Technology, Melbourne, February 11.

Hoffman, R. R., Shadbolt, N. R., Burton, A. M., & Klein, G. (1995). Eliciting knowledge from experts: A methodological analysis. *Organizational Behavior and Human Decision Processes, 62,* 129–158.

Hutchins, E. (1995). *Cognition in the wild.* Cambridge: MIT.

Hutchins, E. (1995). How a cockpit remembers its speeds. *Cognitive Science, 19,* 265–288.

Johnson, A. M. Jr., & Malek, M. (1988). Survey of software tools for evaluating reliability, availability, and serviceability. *ACM Computing Surveys, 20*(4), 227–269 (December).

LaFrance, M. (1987). The knowledge acquisition grid: A method for training knowledge engineers. *International Journal of Man-Machine Studies, 26,* 245–255.

Lahlou, S. (2000). Attracteurs cognitifs et travail de bureau. *Intellectica: revue de l'Association pour la Recherche Cognitive* (Vol. 30, pp. 75–113). ISSN 0769-4113.

Lahlou, S. (2005). Cognitive attractors and activity-based design: Augmented meeting rooms. *Human computer interaction international.* Las Vegas.

Laprie, J. C. (1992). *Dependable concepts: Basic concepts and terminology.* New York: Springer Verlag.

Laprie, J. C. (1994). How much is safety worth? *Proceedings of the IFIP 13th World Computer Congress, III*(A-53), 251–253.

Learmount, D. (1995). Lessons from the cockpit. Flight International. 11/01/95

Lewis, E. E. (1996). *Introduction to reliability engineering.* New York: Wiley.

Loine, L. (2011). *The complexity of management and control of a nuclear power plant.* Lecture given at Florida Institute of Technology, Melbourne, March 18.

Mankins, J. C. (1995). Technology readiness levels. White paper. Advanced Concepts Office, Office of Space Access and Technology, NASA.

Millot, P. (2011). *Human and machine cooperation in process control.* Lecture given at Florida Institute of Technology, Melbourne, April 1.

Morel, M. (2008). *Attention sharing between driving and communication: Towards an ecological and exprimental pluridisciplany approach.* Ph.D. Thesis, University of Lyon, France.

Nielsen, J. (1993). *Usability engineering.* San Diego: Academic.

Nilsen, T., & Aven, T. (2003). Models and Model Uncertainty in the Context of Risk Analysis. *Reliability Engineering & System Safety, 79,* 309–317.

Norman, D. A. (1998). *The invisible computer.* Why good products can fail, the personal computer is so complex, and information appliances are the solution. Cambridge: The MIT. ISBN-10:âL¨0-262-14065–9âL¨ISBN-13.

NTSB. (2010). http://www.ntsb.gov/pressrel/2010/100318.html. Accessed 22 Dec 2011.

Pagés, A., & Gondran, M. (1986). *System reliability evaluation and prediction in engineering.* New York: Springer.

Perrow, C. (1999). *Living with high-risk technologies.* Princeton: Princeton University Press.

Petroski, H. (1992). *To engineer is human: The role of failure in successful design.* Vintage: Random House, ISBN 9780679734161.

Pinet, J. (2011). Traitement de situations inattendues d'extrê me urgence en vol: Test d'un modèle cognitif auprès de pilotes experts (Processing unexpected situations of extreme urgency in flight: Test of a cognitive model with expert pilots). Ph.D. Thesis, Doctoral School of Aeronautics and Astronautics, University of Toulouse, France.

Prasad, D., McDermid, J., & Wand, I. (1996). Dependability terminology: Similarities and differences. *Aerospace and Electronic Systems Magazine, IEEE, 11*(1), 14–21 (Jan 1996).

Prichett, A. (2011). *Aiding the human contribution to aviation safety.* Lecture given at Flroida Institute of Technology, Melbourne, April 8.

Ramu, J. P. (2008). Ph.D. Thesis. *Contextual operational documentation effectiveness on airline pilots' performance.* Toulouse: ISAE.

Ramu, J. P., Barnard, Y., Payeur, F., & Larroque, P. (2004). Contextualized operational documentation in aviation. In D. de Waard, K. A. Brookhuis & C. M. Weikert (Eds.), *Human factors in design* (pp. 1–12). Maastricht: Shaker.

Rasmussen, J. (1986). *Information Processing and Human-Machine Interaction—An Approach to Cognitive Engineering*. Amsterdam: North Holland.

Rauzy, A. (2008). Guarded transition systems: A new States/Events formalism for reliability studies. *Journal of Risk and Reliability, 222*(4), 495-505.

Scapin, D. L., & Bastien, J. M. C. (1997). Ergonomic criteria for evaluating the ergonomic quality of interactive systems. *Behaviour & Information Technology, 16,* 220–231.

Shea, V. (1994). *Net etiquette*. San Francisco: Albion Books.

Shiffrin, R. M., & Schneider, W. (1977). Controlled and automatic human information processing. II. Perceptual learning, automatic attending and a general theory. *Psychological Review, 84*(2), 127–190.

Sutcliffe, A. (1997). Task-related information analysis. *International Journal of Human-Computer Studies, 47,* 223–257.

Swain A. D., & Guttman, H. E. (1983). *Handbook of human reliability analysis with emphasis on nuclear power plant applications* (NUREG/CR-1278, SAND800 200, RX, AN). Albuquerque: Sandia National Laboratories, (August).

Taleb, N. N. (2007). *The black swan: The impact of the highly improbable*. New York: Penguin. ISBN 978-1-4000-6351-2.

UC Atlas (2012). http://ucatlas.ucsc.edu/spend.php. Accessed 2 May 2012.

Vicente, K. J. (1997). Heeding the legacy of Meister, Brunswick, and Gibson: Toward a broader view of human factors research. *Human Factors, 39,* 323–328.

Wears, R. (2011). *The healthcare industry as a life critical system*. Lecture given at Florida Institute of Technology, Melbourne, Feburary 25.

Wikipedia (2011). http://en.wikipedia.org/wiki/Safety-critical. Accessed 12 Feb 2012.

Woodson, W. E., & Conover, D. W. (1973). *Human engineering guide for equipment designers*. Berkeley: University of California Press.

References

Chapter 5
The Making of Complex Systems

Introduction

In human factors, we learned during the last 3 decades that people commit various kinds of errors during operations of life-critical systems, and these errors are the cause of a majority of accidents. Therefore, we developed compensation responses for technology, organizations and people, including various kinds of defenses and resilient strategies, because relevant LCS properties were identified. It seems the problem is much deeper than the shallow solutions that have been found so far. Safety culture has to be based on a long-term educational process that should start at school from an early age. Today, such culture has to be reformatted into a complexity culture. What is a complex system? Why such a system could lead to catastrophic situations?

The natural world is full of complex systems, and scientists have studied such complexity for a long time, trying to identify persistent behavioral patterns, generic phenomena, relationships among components and so on (Mitchell 2009). Today, technological developments result in new kinds of complexity. Scientists developed **models** in order to deconstruct complexity, or more precisely see what was supposed to be complex in a different "simpler" way. For example, a long time ago people used to watch the sky as a complex set of stars. Copernic, Kepler and Galilée deconstructed such complexity by providing models based on two simple variables (i.e., distance and mass). Similarly, today's scientists develop models in order to describe what they observe (i.e., they rationalize complex datasets).

This book focuses on artifact design and use, referred to today as "Human-Centered Design". **Artifacts** (i.e., artificial things) differ from natural things because people build them. They can be simple, but we make them more complex by adding more properties. They can also be complex from the start because they are built to solve complex problems. Complex systems are usually defined as an integrated set of interconnected components. These systems can be aircraft, spacecraft, nuclear power plants, medical operating rooms, disaster-management centers and so on. These systems are inherently complex because they are cognitive prostheses (Ford et al. 1997). An aircraft is a cognitive prosthesis because it has been designed to supply people with flying capacity (remember that people do not fly naturally). We

G. A. Boy, *Orchestrating Human-Centered Design*,
DOI 10.1007/978-1-4471-4339-0_5, © Springer-Verlag London 2013

Fig. 5.1 Clement Ader's
Eole. (Image is in the Public
Domain, copyright free)

actually deconstructed the complexity of flying when we understood that flying was a matter of thrust and lift. Once humans had the right structural devices to ensure lift and the right propulsion to ensure thrust, we managed to fly. **Clement Ader** was one of the first to make this theory concrete. Ader built the first flying machine *Éole* (Fig. 5.1) that he attempted to fly on 9 October 1890 in the vicinity of Paris. He flew 50 m reaching a height of 20 cm, 13 years before the Wright Brothers (ASA 1994). Ader deconstructed the complexity of human flight by proving it was possible to physically stay in the air while propelled by an engine. Later on many other people designed, developed and used aircraft along these lines, refining the manned-flight concept. They mastered complexity of the manned-flight concept both functionally and structurally. Note that we incrementally learned from experience. Also, it is important to mention that aircraft improved over the years in the issues to safety, efficiency and comfort because these important topics were developed symbiotically with pilots. More importantly, the experimental test pilot (ETP) job emerged from a human-centered design approach, even if not formalized as it is today.

For a long time, systems were designed individually without much attention to their coupling to other systems, most importantly people. Their integration into an organizational environment was always the source of various surprises and discoveries, sometimes a few catastrophes. Adjustments were always necessary and often systems had to be redesigned to fit the reality of their actual use. This kind of process seems natural and acceptable during development phases, but is not acceptable once a system is delivered. This is a question of technological maturity that deals with intrinsic complexity, and maturity of practice that deals with extrinsic complexity. This distinction between intrinsic complexity and extrinsic complexity is important. **Intrinsic complexity** results from system architecture and internal relationships among its components. **Extrinsic complexity** results from the activity of the various agents involved in the use of the system (i.e., interaction between the system and its environment including its users). Of course, intrinsic complexity and extrinsic complexity are intimately related, but it is often much easier to study them separately and later investigate how they relate.

Perception of complexity is often a matter of **ignorance**. The more we know about something, the more we find it simple to understand and manipulate. For example, it might be very complex to find your way driving to a place that you never went before. You need a map, directions and most importantly you need to think and act appropriately. After you have gone to a place once, it becomes simpler to get there a second time. You have constructed a pattern (or model) for going to this place. You even refine it by optimizing time and distance. In fact, complexity was mainly in your ignorance of salient entities and relationships among them, which determine the relevant patterns. Discovering and learning patterns/models tends to deconstruct complexity.

In Chap. 2, we already saw that maturity of a product is strongly related to the quality of its high-level requirements. Initial conditions matter when we deal with complex systems. The famous meteorologist Edward Lorenz proposed a weather model that demonstrated we cannot predict precise weather forecasts more than a few days ahead, because of sensitivity to initial conditions (this sensitivity is called the Butterfly effect), but we can predict future weather patterns that are called chaotic attractors (Lorenz 1963). Let us take the 2011 Fukushima Daiichi disaster as an example.

The Fukushima Daiichi disaster was first a huge earthquake and tsunami that killed 20,000 people, and a subsequent nuclear catastrophe. Could this event have been predicted? Certainly not in the usual sense of linear mathematical prediction! However, there is geological evidence of six catastrophic tsunamis hitting the Sanriku coast within 6,000 years.[1] Among them, the 1896 Meiji Sanriku earthquake caused 22,000 casualties. Does this knowledge enable prediction? It was very difficult to predict when another earthquake/tsunami would happen when the decision was made to build a nuclear power plant in Fukushima Daiichi. However, it was clear that (1) the risk was very high due to the fact that this area of the world is very vulnerable regarding high-magnitude earthquakes and tsunamis, and (2) if an earthquake and/or tsunami did occur, there would be disastrous consequences. In addition, several decisions were made that did not take into account these two factors. Seismology deals with highly complex geological behaviors that are almost impossible to predict in terms of time of occurrence. However, there are highly possible areas of danger that can be considered as geographical attractors.

Back to design and engineering, following Lorenz's complexity approach, an end product can be very sensitive to high-level requirements since the processes and organization that contribute to the making of the product are complex. However, it would be possible to predict relevant attractors that will shape the main characteristics of the product. For example, it is now clear that taking a technology-centered approach will lead to the development of user interfaces that attempt to compensate and ultimately hide intrinsic complexity, and sometimes force users to adapt to the system (the inside-out approach to engineering). Conversely, taking a human-centered approach based on extrinsic complexity from the start will lead to an integrated symbiotic human-machine system in the end (the outside-in approach to design).

The inside-out approach has been used for a long time because engineered systems were less complex than the ones we know today. Therefore, it was easy in the past to design a user interface because the number of variables and parameters was reasonably small. Today, this number has become huge, and designers have to make difficult choices in the design of user interfaces. What is important to show? What should be controlled? Layers and layers of software have been developed to take into account safety, efficiency and comfort at the same time, increasing both intrinsic and extrinsic complexity. Consequently, systems have their own behaviors that people need to perceive, understand, consider and react to appropriately. We have moved into a human-system interactive world that cannot be considered

---

[1] http://en.wikipedia.org/wiki/Seismicity_of_the_Sanriku_coast.

as a single human operator using a machine, but as a multi-agent interactive environment. This is why the classical **positivist approach**[2] is no longer sufficient, often ineffective and inappropriate, so we prefer to use more **phenomenological approaches** to design and development. The positivism-phenomenology distinction will be emphasized in this chapter. In particular, a complex system is an articulated set of dedicated sub-systems that induces emerging phenomena, typically unknown at design time. Good human-centered design should focus on discovery of emerging phenomena.

Therefore, the **systems-of-systems** (SoS) approach is preferred to the commonly-used approach of designing systems individually and then integrating them in the end. Without designing the overall architecture and enabling its functionalities and behaviors, it is not possible to assess its extrinsic complexity (nor its intrinsic complexity). Various definitions of the SoS concept have been offered by many authors (Jamshidi 2005). It requires more work to get a workable definition. It originated in the defense sector (Luzeau and Ruault 2008). Today, we cannot think of air traffic management of the future without stating it in terms of systems of systems. Other examples are the Internet, intelligent transport systems, and enterprise information networks (or Intranets). It becomes obvious that this notion of systems of systems now integrates the distinction of intrinsic and extrinsic complexity.

Since systems of systems are made of humans and systems, original natural systems tend to become artificial systems. Many examples have emerged from this evolution such as genetically modified organisms, integrated prostheses, and so on. Technology is now part of our lives and needs to be investigated correctly if we want to take care of Human Kind and the Earth. This is why improving our understanding of complexity is very important when we, Humans, attempt to modify natural complexity. Consequently, **life-critical systems** need to be better investigated and understood in order to find the right mix between humans and machines.

Non-linearity, Attractors and Chaos

A dynamical system can be represented by a state vector, $x = (x_1, \ldots, x_n)$, time t, and an evolution function f that transforms a state at one time to another state at some other time:

---

[2] The French philosopher and sociologist Auguste Comte introduced positivism in the beginning of the nineteenth century. Positivism asserts that the only authentic knowledge is that which is based on sense experience and positive verification. The German philosopher Edmond Husserl introduced phenomenology in the beginning of the twentieth century as the study of consciousness and conscious experience. Among the most important processes studied by phenomenology are intentionality, intuition, evidence, empathy, and intersubjectivity. The positivism-phenomenology distinction opens the debate on objectivity and subjectivity. Our occidental world based most of our design and engineering on positivism which led to developing a very precise and verifiable syntax, often leaving semantics somewhere behind, perhaps because semantics is full of subjectivity. It is time to re-qualify phenomenology in design and engineering. A few organizations and companies already work in this direction.

$$\frac{dx}{dt} = f(x, t, u, p, q),$$

where $u = (u_1, \ldots, u_m)$ is a control vector, $p = (p_1, \ldots, p_k)$ is a vector of system parameters, and $q = (q_1, \ldots, q_l)$ is a perturbation vector.

Ordinary differential equations are commonly used to model and simulate dynamical systems. Other mathematical approaches include finite state machines, cellular automata, Turing machines, stochastic equations and partial differential equations. When systems can be modeled by these approaches, the temporal behavior of such systems can be analyzed.

There is no universal mathematical tool that enables handling non-linear dynamical systems in their complete form by analogy with, for example, linear systems and linear algebra. It is important to note that Henri Poincaré proposed a number of general ideas and rules that could help scientists choose and implement the corresponding mathematical techniques to solve specific problems in Mathematics. Poincaré put together the qualitative theory of differential equations. Poincaré's findings helped the analysis of non-linear dynamical system using differential equation taking into account their evolutions and critical elements such as limit cycles, attractors, chaos, bifurcations and singularities. I advise the reader to refer to the Santa Fe Institute[3] of complexity, that has already produced excellent material on complexity and non-linear dynamical systems, and other work such as (Mitchell 2008; Holland 1998).

Understanding complexity is difficult. Understanding that a problem may have more than one solution is a first start. Understanding that two slightly different initial conditions may lead to drastically different final results is a second step. Understanding that a small variation in an input may result in huge consequences in non-linear dynamical systems is a third step, and so on. Understanding that the behavior of a complex system cannot be fully predictable can be very negative, but understanding that a complex system has some specific properties is very positive. This is precisely what Poincaré understood (i.e., some complex equations cannot lead to meaningful quantitative solutions, they can lead to informative qualitative solutions or properties). This is why we need to look for interesting properties (or attractors) of complex systems. Lorenz's non-linear dynamical equations of his toy atmosphere led to what he called the butterfly effect, a beautiful attractor (Lorenz 1963).

An **attractor** can be seen as an envelope that "includes" all trajectories of a non-linear dynamical system generated from different initial conditions. Each trajectory, or data set, as related to a specific initial condition is an unpredictable instance. Conversely, the accumulation of several trajectories of a non-linear deterministic system defines an attractor that can be anticipated once it is discovered. Chaos theory is mainly based on these premises. More specifically, chaos theory studies the behavior of dynamic systems that are highly sensitive to initial conditions. An attractor can therefore be also defined as a concentration of chaotic trajectories of a non-linear dynamical system.

---

[3] The Santa Fe Institute is a private, non-profit, multidisciplinary research and education center, funded in 1994. It is primarily devoted to basic research.

Maturity of a complex system cannot be fully reached until behavioral attractors are identified. Chaos theory provides theoretical models such as strange attractors and the butterfly effect (Poincaré 1890; Lorenz 1963). Systems of systems are embedded into each other, sometimes repeating similar structures at a different level of granularity. This is why Fractal theory can be very useful to better analyze and understand both their structure and dynamics (Mandelbrot 2004). Their multiple interactions manage to develop emergent behaviors that are useful to identify. Finally, emergence cannot be sustained without self-organization (Ashby 1947). Socio-cognitive stability, whether passive or active, is an emergent property of a system of systems. Knowing such properties is crucial because self-organization as a complex system does not require any central authority, as long as organization and coordination rules are well defined and used. Think about a huge flock of birds describing patterns in the sky. Their organization and coordination rules ensure the stability of the whole flock. Reynolds proposed a model for flocking that includes three properties (Reynolds 1987): alignment that consists in moving in the same direction as neighbors; separation and collision avoidance that consists of short-range repulsion; and cohesion that consists of remaining close to each other or long-range attraction.

Natural Versus Artificial Complexity

First, let us make a major distinction between natural and artificial complexity. Natural complexity typically denotes complexity of natural systems such as vegetal and animal living organisms, and geological systems. In contrast, artificial complexity denotes complexity of human-made systems (or artifacts), such as mechanical systems.

Natural complexity is incrementally explored from the outside-in. For example, chemists describe plants using attributes that characterize constant behaviors and processes. They try to discover properties. Exploration is a major process used to uncover natural complexity. Research consists of exploring what the major agents involved in the natural systems being investigated are, whether a living organism or a geological system. Each agent exhibits a behavior that must be identified, interpreted and explained. Agents are usually interconnected among each other and complexity can be explained in terms of interconnectivity. Typically, a natural system cannot be easily decomposed because some connections are crucial for the life and/or integrity of the overall system. However, decomposition and categorization are very important processes commonly used in research. For example, a human body can be decomposed into organs, bones, blood and so on. The main issue is that these components are difficult to study without taking into account the various interconnections between them. This issue is called **separability** (e.g., we cannot study the human heart in isolation since it is necessarily interconnected with other organs in order to work properly on all levels). Although, when the right variables and models are identified, it may become possible to isolate some parts and study them in isolation (e.g., biologists have been trying to learn about cells by culturing

Fig. 5.2 Old clock
mechanics. (Stock image
courtesy of
iStockphoto/Thinkstock)

them in isolation). The closed biological system Biosphere 2,[4] for example, is a 3.14 acre research facility owned by the University of Arizona that is intended to simulate living systems in order to better understand Earth and its place in the universe.

Separability, or non-separability, has already been studied in mathematics, physics, chemistry and physiology. For example, in quantum mechanics, the Schrödinger equation describes how the quantum state of a physical system changes over time. There are some cases in which this equation can be separated into several ordinary differential equations that simplify the resolution of the mathematical system (Eisenhart 1948). In chemistry, a mixture of substances can be transformed into several distinct products through a separation process. Separability needs to be analyzed in conjunction with contextuality (or locality). Claude Bernard, a French physiologist and surgeon, discovered the concept of "milieu intérieur" (eBooks-France, Bernard 2000), which was later coined by Walter Cannon as "homeostasis". A complex system such as the human body consists of a huge number of feedback loops among various components that cannot be partly broken for the health of the whole, and some of them cannot be broken at all, otherwise the whole would die.

Artificial complexity can formally be identified and explored before a system is made. However, we also need to make other distinctions between **hardware complexity** and **software complexity**. Automata have been built for a long time. For example, the clock for example has existed for ages, but for a long time its complexity was only mechanical (hardware complexity).

Mechanical complexity is tractable, decomposable, and linearizable. For example, old-clock mechanics could be decomposed into a set of articulated mechanisms intimately organized (Fig. 5.2). It was not easy to make such clocks work because manual and mental skills were required. However, the complexity resulting from clock design could be tracked (e.g., clock mechanisms could be put on the table and related to each other in a linear way). They could be repaired when they failed. Conversely, modern watches are electronic and software-intensive. Their maintenance does not require any specific skills in mechanics as in the past, since the time system

---

[4] http://www.b2science.org/about/fact.

Fig. 5.3 Enjoying the complexity of wine! (Stock image courtesy of iStockphoto/Thinkstock)

is a computer program. In contrast, when they fail, the entire software system needs to be changed. Usually they are not repaired, we simply change the watch.

People can transform natural entities to make new varieties of objects. Wine for example is such a generic object. It is made from grapes through fermentation and subtle chemical transformations. The beauty of wine is that it may have various kinds of behavior with respect to the terrain where grapes are cultivated, the evolution of weather during the year and so on. Wine is a complex system, both natural and artificial.

In order to analyze the complexity of a good wine, for example, experts could say something like "a combination of richness, depth, flavor intensity, focus, balance, harmony and finesse" (Fig. 5.3). This pattern can be called the "signature" of the wine being tasted. In chaos theory, we would talk about an attractor. For example, Bordeaux wines, while they can be very different among each other (i.e., they have different "trajectories"), they all are consistent with the same signature (i.e., Bordeaux wines' attractor is very distinguishable). A Bourgogne wine has a different signature for example.

Wine complexity can be described by words such as[5] "acetic, acidic, ageworthy, aggressive, ample, aromatic, astringent, austere, balanced, big, bitter, blockbuster, body, bold, bouquet, buttery, bright, character, clean, complex, concentrated, cooked, corked, crisp, deep, dry, dull, easy-drinking, elegant, fat, flabby, fleshy, focused, fresh, fruity, grassy, green, hard, harsh, heavy, herbal, honeyed, jammy,

---

[5] http://world-food-and-wine.com/describing-wine.

lean, light, long, madeirized, mature, meaty, mineral, neutral, noble, nose, nutty, oaky, oxidized, petrol, piercing, powerful, racy, rich, ripe, rounded, simple, smooth, soft, sparkling, spicy, steel, stony, structured, subtle, sulphurized, supple, sweet, tannic, tart, toasty, velvety, warm, woody, yeasty… and so on". These words attempt to describe the terminology of wine tasting. Of course, such terminology would become a real ontology if a wine expert used it, someone who knows how to relate the various terms with the interconnected concepts that characterize wine. In particular, experts know that such terms are useful to denote emerging properties resulting from the various relevant transformations and evolutions (of wine in this case). Again, complexity is in the number of independent terms and concepts being used, as well as the interconnections between them.

Today, technology develops very fast and is increasingly interconnected. Not only are artificial systems interconnected, natural and artificial systems are interconnected as well. This is why human-centered design has become tremendously important. Making a good wine requires knowledge, knowhow, experience and risk taking. Wine making can qualify as a **system of systems** (see a workable definition later in this paper). Wine emerges from mechanical and chemical transformations of grapes. These transformations can be considered as sub-systems. In the same way, developing a highly interconnected complex system requires knowledge, knowhow, experience and risk taking. For example, current and future air traffic management systems are systems of systems where there are artificial systems (e.g., onboard and ground systems) and humans (e.g., pilots, air traffic controllers, airworthiness officers, etc.). Such a system of systems evolves where both artificial and natural (sub)-system behaviors are incrementally transformed and adapted to ensure consistency of the whole. It is very difficult to make a good wine because the transformation and adaptation of each agent can be challenging to master. Wine makers need to know how wine agents interact amongst each other. In the same way, in the ATM we need to incrementally understand how ATM agents interact among each other, taking into account that as the number of aircraft increases, the nature of their interconnections evolves.

Toward a Socio-Cognitive Framework

Multi-agent systems have been studied for a long time in artificial intelligence (Ferber et al. 2009; Nair et al. 2003; Ferber 1999; Bradshaw 1997; Minsky 1985). In multi-agent modeling, the first difficulty is to define the various relevant agents, and second their interdependence and relationships, as well as how they interact with each other. Note that in this chapter, I do not want to limit the description of agents to machines or what is now called intelligent (artificial) agents; agents can be natural and artificial. We also need to be able to observe them in the real world in order to acquire their behavior and later on identify their internal mechanisms. They all behave with respect to their intentions (goal-driven behavior and functions) and reactions (event-driven behavior and functions).

Identifying **interaction patterns** among agents and finding out their internal mechanisms is actually an outside–in approach. Internal mechanisms of both artificial and natural agents are algorithms supported by appropriate internal architectures motivated by observed an/or desirable interactions. Deduced algorithms and architectures of artificial agents are very useful requirements for the design of real systems. Deduced algorithms and architectures of natural agents (typically people) are attributes for determining how they might/should train and work. This is why modeling and simulation of such multi-agent systems are so important very early during the design process of SoSs, to **identify the appropriate interaction space made of appropriate resources and contexts**.

At this point, it is important to make clear that the outside-in approach to making complex systems needs to be supported by a solid framework. The cognitive function framework is one of them (Boy 1998). When a cognitive function is allocated to an agent, it provides a specific role to this agent. If the agent is a postman, his or her main cognitive function (i.e., his or her role), is to deliver the mail. In addition, the postman belongs to the mail services agency (i.e., a system of systems). A cognitive function is also defined within a context (e.g., the postman delivers letters from 8 a.m. to 12 p.m. and 2 p.m. to 5 p.m. every weekday in a well-defined neighborhood). The cognitive function context can then be temporal or spatial, but it also can be normal or abnormal. For example, a strike could be an abnormal context when the postman in duty has longer working hours and a bigger neighborhood. Finally, a cognitive function has resources that support its achievement (e.g., the postman has a bag, a bicycle and a special mental skills to pattern-match addresses on envelopes to street names, house numbers and people's names). The formers are physical resources and the latter (i.e., pattern-matching) is a cognitive resource that is itself a cognitive function. Consequently, a cognitive function can also be a "society of cognitive functions." In some abnormal contexts, such as a strike, postmen on duty cannot do the job by themselves, and need to delegate part of the delivery task to several other people. These people may not be trained and need to be supervised. Postmen on duty need to have cognitive functions such as "training other people to deliver letters," "supervising," "evaluating performance," and so on. Note that the "letter delivery" cognitive function can be decomposed into several other cognitive functions that themselves can be distributed among a set of agents.

The cognitive function framework is structured around two spaces: the context space and the resource space (Boy 2011).

In **context space**, the main entity is a **procedural scenario** or chronological script. A procedural scenario can be described as a tree that may have normal and abnormal branches. The basic entity that is commonly used to develop procedural scenarios is the interaction block or i-Block (Boy 1998). The central component of an i-Block is its context C. The context of an i-Block is a set of (persistent) conditions that need to be satisfied to enable the i-Block. i-Blocks are mutually inclusive (i.e., a context of i-Block is an i-Block). i-Blocks are procedurally organized (i.e., an i-Block is necessarily followed by another i-Block except when it terminates a context). A terminating i-Block has a goal (or an abnormal condition) that is the same as the goal (or an abnormal condition) of its context. Similarly, a starting i-Block has a triggering condition that is the same as the triggering condition of its context.

The triggering condition TC of an i-Block is a set of conditions that activate the i-Block when they are satisfied and the i-Block is enabled. The goal G of an i-Block is a set of conditions that terminates the activity of an i-Block when they are satisfied. An abnormal condition AC of an i-Block is a set of conditions that terminates the activity of an i-Block when they are satisfied. Therefore, the activity of an i-Block can be terminated either normally when its goal is reached or when an abnormal condition is satisfied.

Therefore, an i-Block can be defined as: iB = {TC, G, AC, C, CF}, where CF is the cognitive function that enables to execute the task assigned to the i-Block). In fact, an iB is the execution of a CF in a given context C. TC, G and AC enable the connection between several iBs.

In the **resource space**, the main entity is the **declarative scenario** or organizational configuration. A declarative scenario can be described by a network of cognitive functions and physical artifacts. The basic entity that is commonly used to develop procedural scenarios is the cognitive function or CF (Boy 1998). A CF can be described by three attributes: its role R, its context of validity C and its resources $\{RES_i\}$. C can be normal C_N or abnormal C_A. A resource can be a cognitive CF or a physical artifact PA. Therefore, a cognitive function can be defined as: CF = {R, C, $\{CF_j, PA_k\}$}. In addition, there is a relationship between cognitive functions and agents (i.e., the function allocation issue). Agents have competencies that enable the allocation of appropriate functions whether they are cognitive or physical. We will call this allocation relation Alloc (CF, A), where CF is a cognitive function and A an agent. Therefore, functionally speaking, an agent can be defined by a set of cognitive functions (describing her/his/its competence). Of course, as we already saw in the postman example, cognitive functions can be related to each other (cognitive function network) independently of their allocation to agents. Consequently, there will be a set of constraints that will define possible allocations among agents.

Cognitive functions are not only allocated deliberately, they may also emerge from interactions among agents as necessity. This phenomenon of emergence needs to be captured, analyzed and rationalized (Cognitive Function Analysis, see Boy 1998). Emergent cognitive functions are incrementally added to the cognitive function network together with their allocation to appropriate agents.

Some scenarios, whether procedural, declarative or both, can be generic and can be reused in other scenarios. An HCD architect knows many generic scenarios and is able to detect very quickly what clients want and need. This is why research is needed to develop such generic scenarios in order to use them in human-centered design.

Intrinsic Versus Extrinsic Complexity: The Issues of Maturity

As an example, cars are now equipped with new artificial agents, such as:

- cruise control system that maintains constant speed;
- Anti-lock Braking Systems (ABS), Traction Control Systems (TCS, TRC, ASR), Electronic/Dynamic Stability Control Systems (ESP, ASR, DSC) that keep vehicles on track when surface road conditions are not favorable or help maneuvering with vehicle in off-nominal driving situations and conditions;

- Global Positioning System (GPS) that assists a driver in navigation tasks;
- Line Keeping Assistant (LKA) that enables the car to automatically follow a line on the road;
- collision avoidance system that enables a driver to know how close the car is to other cars surrounding it;
- hand-free telephone kit that enables a driver to communicate with other people outside the car, and so on.

Each system usually supplies a specific function that deals with safety, efficiency and comfort: speed control, navigation assistance, trajectory guidance, collision avoidance, and communication support. Each of these functions are great individually because they are solutions to individual problems, and their individual intrinsic complexity is typically mastered. However in some situations, when these functions are collectively used, they may induce high workload, situation awareness and decision-making issues, as well as action taking problems. Drivers need to share their attention among several artificial agents (i.e., systems) that are not integrated (i.e., drivers have new function allocation tasks to perform in order to manage resulting uncoordinated multi-agent systems). We will say that extrinsic complexity is not mastered. This kind of complexity can be qualified as **human-machine coordination complexity**. Such complexity can be represented and tested using the CF-based socio-cognitive framework presented above.

We now understand that extrinsic complexity as a matter of function allocation among agents. If we see modern car cockpits as systems of systems, there is a need for better understanding how the various systems are interconnected not only at design time but also at use time. In other words, it is not sufficient to verify that systems work well intrinsically, they need to be used safely, efficiently and comfortably. For that matter, it may take some time to understand how the various functions have to be allocated, interconnected and coordinated within the context of driving, and more generally during operations. This logically leads to technology maturity (i.e., technology reliability and robustness) and maturity of practice (i.e., interaction among agents).

Maturity is often difficult to measure because technology constantly changes. The problem is to find out how technology changes; it can be an **evolution** or a **revolution**. Technology evolves when small increments are implemented, and practice changes slightly. A technology revolution is observed whenever jobs that it induces drastically change. When systems (e.g., GPS and LKA) are incrementally introduced in an existing system (e.g., a car), this is usually perceived as an evolution. However, when they become too numerous, the job of the user drastically changes because it is no longer possible to manage these systems without a careful and tedious dynamic function allocation. In addition, the overall situation may become dangerous when the resulting function allocation process is too demanding in real-time. Typically, systems are incrementally added and accumulated until behavior of the overall system of systems drastically changes. At this point, a totally new system emerges and we observe a revolution. New strategies need to be found to handle the emergent system, and eventually redesign it.

What we described above is a matter of SoS **configuration.** However, together with the structure there is function that enables behavior. Complex systems also have behavioral properties such as attraction, fractality, emergence and self-organization. From a complexity science point of view, emergent properties could be seen as behavioral attractors.

Dealing with New Types of Complexity and Uncertainty

Our technological world is always changing. Therefore, people are required to adapt to new types of complexity continuously due to the change of cognitive functions that are statically allocated by the introduction of new technology, and of cognitive functions that dynamically emerge from the use of this technology. A number of complexity-related concepts were elicited from experts in the aviation domain, using appropriate knowledge acquisition methods (Boy 2007). The major concept that emerged was "novelty complexity. "

Complexity and Maturity

Many people interact with a computer everyday without caring about its internal complexity... fortunately! This was not the case barely 30 years ago. Individuals who were using a computer needed to know about its internal complexity from both architectural and software points of view, in order to make it do the simplest things. Computer technology was not as mature as it is today. Computer users had to be programmers. Today, almost everyone can use a computer. However, internal complexity may perhaps become an issue in abnormal or emergency situations.

Internal complexity is about *technology maturity*. Maturity is a very complex matter that deals with the state of evolution of the technology involved, and especially reliability. Are both the finished product and related technology stabilized? Internal complexity of artifacts that are mature and reliable (i.e., available with an extremely low probability of failure), is or not at all perceived. In other words, when you can delegate with confidence and the work performed is successful, the complexity of the delegate is not an issue. However, when it fails, you start to investigate why. You look into the "black-box. " For that matter, the "black-box " should become more understandable (i.e., an appropriate level of complexity must be shown to the user). This implies the user should be able to interpret this complexity. Therefore, either the user is expert enough to handle it or needs to ask for external help. For example, current sophisticated cars and trucks are so computerized that when something is suddenly goes wrong, the driver is typically not able to understand the situation and consequently comes to a decision, perhaps not acting inappropriately. Particular kinds of purposeful information should be provided to avoid even the worst consequences. Various levels of explanations should be available according to context; this is a difficult thing to do.

Perceived complexity is about *practice maturity*. Is this technology adequate for its required use, and how and why do, or don't, users accommodate to and appropriate the technology? Answers to this question contribute to a better understanding of perceived complexity, and further development of appropriate empirical criteria. Perceived complexity is a matter of the relationship between users and technology. Interaction with an artifact is perceived as complex when the user cannot do or has difficulty doing what he or she decides to do with it. Note that users of a new artifact still have to deal with its reliability and availability, which are not only technological, but are also related to tasks, users, organizations and situations. This is why novelty complexity analysis and evaluation require solid structuring into appropriate categories.

A user who interacts with a complex system inevitably builds expertise. He or she cannot interact efficiently with such an artifact as a naive user. There is an adaptation period for new complex tasks and artifacts because complex artifacts such as airplanes are prostheses. The use of such prostheses requires two major kinds of adaptation: mastering capacities that we did not have before using them (e.g., flying or interacting with anyone anywhere anytime); and measuring the possible outcomes of their use, mainly in social terms. All these elements are intertwined, involving responsibility, control and risk/life management. Therefore, provided evaluation criteria cannot be used by just anyone. These criteria must be used by a team of human-centered designers who understand human adaptation. You may ask, who could disagree with this? Today, the straight answer is, the finance-driven decision-makers, just because it is too expensive in the short-term. But, how can we manage maturity with short-term goals?

Maturity Management

It is expected that the complexity of an artifact varies during its life cycle. Therefore, both technology and practice maturities need to be taken into account along the life-cycle axis of a product by all appropriate actors; I call "maturity axis" a sequence of maturity checkpoints and re-design processes up to the entry into service. There are methods that were developed and extensively used to improve the efficiency of software production processes such as Capacity Maturity Model Integration or CMMi (Paulk et al. 1993). CMMi partly contributes to assure technology maturity, in the sense of quality assurance. However, this method does not address directly either internal complexity or perceived complexity of the artifact being developed. This is why maturity checkpoints that involve usability (Nielsen 1993) tests during the whole life cycle are strongly recommended.

"Complexity refers to the internal workings of the system, difficulty to the interface provided to the user—the factors that affect ease of use. The history of technology demonstrates that the way to provide simpler, less difficult usage often requires more sophisticated, more intelligent, and more complex internal workings. Do we need intelligent interfaces? I don't think so: The intelligence should be inside, internal to the system. The interface is the visible part of the system, where people need stability, predictability and a coherent system image that they can understand and thereby learn" (Norman 2002). Norman's citation is very important today when we have layers and layers of software piled on top of each other, sometimes designed and developed to correct previous flaws of lower layers. Software engineers commonly talk about patches. This transient way of developing artifacts does not show obvious maturity. Technology maturity, and consequently internal complexity, of an artifact can be defined by its integration, reliability, robustness, socio-cognitive stability and availability. As previously stated, it is always crucial to start with good, high-level requirements that, of course, will be refined along the way. Problems arise when those requirements are weak.

Technology-Centered Internal Complexity Versus User-Perceived Complexity: Focusing on Cognitive Stability

Schlindwein and Ray (2004) introduced the distinction between descriptive and perceived complexity as an epistemological problem of complexity. Perceived complexity may block users when they cannot see or anticipate the outcome of their possible actions (i.e., world states are so intertwined that they cannot see a clear and stable path to act correctly). In addition, they may become aware of such complexity after they commit errors and need to recover from them. At this point, it is appropriate to say that real situation awareness is learned in this kind of error-recovery situation.

Therefore, user-perceived complexity is intimately linked to cognitive stability. Technology provides **cognitive stability** when it is either self-recoverable (like the automated spelling checker that automatically correct typos) or supports users with the necessary means to anticipate, interact and recover by themselves. The main problem with self-recoverable systems is reliability and robustness. When they are not reliable or robust enough, people stop using them. We know that users detect most of their errors, and recover from them when they have the appropriate recovery means, whether these means are technological (i.e., tool-based) or conceptual (i.e., training-based). It was observed that in air traffic control, controllers detected 95 % of their errors (Amalberti and Wioland 1997). Obviously, the required level of the user's experience and expertise may vary according to the cognitive stability of the overall human-machine system.

When a person controls a machine, two main questions arise:

1. Are machine states observable (i.e., are the available outputs necessary and sufficient to figure out what the machine does)?
2. Are machine states controllable (i.e., are the available inputs necessary and sufficient to appropriately influence the overall state of the machine)?

A mental model is developed to control a machine, associating observable states to controllable states (Rasmussen 1986; Norman 1986). There is a compromise to be made between controlling a system through a large set of independent observable states and a small set of integrated observable states. "... The larger the number of degrees of freedom in a system, the more difficult it is to make the system behave as desired (i.e., perceived complexity is higher). Simply counting degrees of freedom, however, oversimplifies the issue. It is the manner in which degrees of freedom interact that determines the difficulty of controlling a system. For example, if the n degrees of freedom of a system are independent of one another, then the controlling system needs only to process an algorithm that is adequate for the control of a single degree of freedom; the algorithm can be replicated n times to control the overall system. Conversely, if the degrees of freedom are dependent (that is, if the effects of specifications of values for a particular degree of freedom depend on the values of other degrees of freedom), then a team of independent controllers is no longer adequate, and more complex control algorithms must be considered. (Jordan and Rosenbaum 1989; Norman 2002).

Cognitive stability is analyzed using the metaphor of stability in physics. Stability can be static or dynamic. Static stability is related to the degrees of freedom (e.g., an object in a three-dimensional world is usually defined by three degrees of freedom). A chair is stable when it has (at least) three legs. Human beings are stable with two legs, but this is a dynamic stability because they have learned to compensate for, often unconsciously, their instability. When an object is disturbed by an external event, there are usually two cases: a case where the object returns to its original position, we say that the object is in a stable state; and a case where the object diverges from its original position, we say that the object is (or was) in an unstable state. When a user acts erroneously, there are two cases: in a case where the user recovers from his or her erroneous action, we say that the user is in a stable state; and in a case where the user does not recover from his or her erroneous action, we say that the user is (or was) in an unstable state.

There are erroneous human actions that may be tolerated, and others that should be blocked. Error tolerance and error resistance systems are usually related to useful redundancy. Error tolerance is always associated with error recovery. There are errors that are acceptable to make because they foster awareness and recovery. However, recovery is often difficult, and sometimes impossible, when appropriate resources are not available. Action reversibility should be put forward and exploited whenever a user can backtrack from an erroneous action, and act correctly. The UNDO function available on most software applications today provides a redundancy to users who detect typos and decide to correct them. Thus, making typos is tolerated, and a recovery resource is available. Error resistance is, or should be, associated to risk. Error-resistance resources are useful in life-critical systems when high risks are possible. They may not be appropriate in low-risk environments because they usually disturb task execution. For example, text processors that provide permanent automatic grammar checking may disturb the main task of generating ideas, and unwanted automatic "corrections" to text that are off the visible screen can go unnoticed. Inappropriate learning and training, poor vigilance, poor feedback, too much interruption, fatigue and high workload are important adverse influences on cognitive stability.

Organizational Complexity: Focusing on Socio-Cognitive Stability

The exploration of organizational automation leads to the investigation of the concept of organizational complexity. The management science literature is certainly rich in definitions of organizational complexity, but they still need to be improved in order to take into account organizational automation. For example, Dooley (2002) defines organizational complexity as "... the amount of differentiation that exists within different elements constituting the organization. This is often operationalized as the number of different professional specializations that exist within the organization. For example, a school would be considered a less complex organization than a hospital, since a hospital requires a large diversity of professional specialties in order to function. Organizational complexity can also be observed via differentiation in structure, authority and locus of control, and attributes of personnel, products, and technologies." In other words, the number and diversity, as well as authority distribution, of agents and, more specifically, cognitive functions involved in the organization are direct contributing factors to organizational complexity.

Luhmann (1995) states "we will call an interconnected collection of elements complex when, because of imminent constraints in the elements' connective capacity, it is no longer possible at any moment to connect every element with every other element... Complexity in this sense means being forced to select; being forced to select means contingency; and contingency means risk". In this definition, connectivity is the problem where channels for communication, cooperation and coordination are crucial. For example, accidents may occur when two different agents use their authority to decide to act in ways that are finally incompatible for the overall product. In fact, these agents were supposed to be independent, but it turned out that their dependency emerged from the situation. Therefore, since emergent cognitive functions are generally situated, they cannot be discovered without operational experience (i.e., either using a simulation facility or in a real-world environment). More specifically, there is an account of two hospital services that were somehow disconnected, and both eventually administered incompatible drugs to the same patient, each of these drugs being totally acceptable independently. Another example can be taken from the aviation domain, where a wiring system was being developed by a division of a large company and was not aware of the integration constraints of another division. In the former case, the lack of connectivity might have caused the death of the patient without the intervention of the patient's family; in the latter case, it caused a drastic delay in the delivery of the product and subsequent financial issues.

The complexity of the product itself has a major influence on the organizational complexity. Designing a large commercial aircraft is not the same as designing an ordinary meal. In order to reduce organizational complexity, a typical strategy of financial management is to divide the overall design and manufacturing work into small pieces that are simple enough to be performed by cheap labor. The main question that remains to be solved is the integration issue that requires both global and specific technology competence. This kind of competence is progressively removed from current industrial organizations and replaced by reporting mechanisms mainly focused on financial factors. Instead, connectivity in the sense of domain-specific, as well as educated common sense articulation, work should be further developed where people would be involved in a participatory way. Instead of relying implicitly on articulation work performed by a few motivated people, it is crucial to focus the way people communicate, cooperate and coordinate within the organization; this is what I call human-centered continuum in the organization. Therefore, the main issue is to (re)-create a human-centered continuum in the organization instead of dichotomized pieces of a financial puzzle, hoping that these pieces will magically connect among each other in the real life. In fact, socio-cognitive stability is enhanced when people deploy a collaborative involvement toward a mature product.

Socio-cognitive stability (SCS) has been defined as local and global SCS (Boy and Grote 2009). Local SCS is defined as the optimum agent's workload, situation awareness, ability to make appropriate decisions and, finally, correct action execution. It can be supported by appropriate redundancies and various kinds of cognitive support such as trends, relevant situational information and possible actions. Global SCS is defined as the appropriateness of functions allocated to agents, pace of information flows and related coordination. It is very similar to the level of synchronization of rhythms in a symphony. Globally, socio-cognitive support could be found in a safety net that would take into account the evolution of interacting agents and propose a constraining safety envelope in real time.

The major problem is the mismatch between the growing number of interdependencies and the lack of concrete links that should materialize these interdependencies. They are supposed to be glued in the end, but this is where the system does not work because the necessary gluing process is often performed in a hurry, at "the last minute," or the gluing requires expert human judgment, which is excluded from the process. Boy and Grote (2009) proposed a measure of socio-cognitive complexity derived from several contributions such as Latour's account on socio-technical stability (Callon 1991; Latour 1987), emerging cognitive functions (Boy 1998), distributed cognition (Hutchins 1987), and socio-cognitive research and engineering (Hemingway 1999; Sharples et al. 2002). Three kinds of measures were deduced during the PAUSA project for assessing socio-cognitive stability:

1. *time pressure criticality* (e.g., in the air traffic management (ATM) case, the amount of workload that an agent (or a group of agents) requires to stabilize an ATM system after a disturbance) is a workload measure that could be assessed as the ratio between the sum of required times for each action on the total available time (Boy 1983);
2. *complexity* could be characterized by the number of relevant aircraft to be managed per appropriate volumetric zone (AVZ) at each time[6];

---

[6] An AVZ is calculated with respect to the type of flow pattern (e.g., aircraft crossing, spacing and merging). The definition of such an appropriate volumetric zone requires the assistance of operational ATC controllers. From a socio-cognitive perspective in ATM, complexity should be considered together with capacity. This is what the COCA (COmplexity & CApacity) project

3. *flexibility* could be characterized by the ease of modification of an air-ground contract in real-time—flexibility assessments should guide ATM human-centered automation and organizational setting.

Metrics representing socio-cognitive stability could be added as a fourth kind of measure. Overall, increasing the number of agents and their interdependencies increases complexity and can increase uncertainty, which need to be managed by finding the right balance between reducing uncertainties through centralized planning and coping with uncertainties through decentralized action. It is very difficult and most often impossible, to achieve overall system goals without loose coupling among actors who are required to be both autonomous and intimately coordinated (Grote 2004).

Positivist Versus Phenomenological Approaches

The difficulty in the making of complex systems is mastering novelty and consequently maturity. Product maturity is a matter of high-level requirements and iterative tests during the various design and development processes (Boy 2005). This is why if you start with good architecture in the first place, you are very likely to end up with what you really want. Conversely, if the starting architecture is not well thought out, you will end up with constant modifications to reach satisfaction. The right dose of expertise and common sense is required to define high-level requirements. Afterwards, intrinsic complexity will be mastered if the right engineers are involved at the right time and in the right configuration. In contrast, extrinsic complexity will be a matter of mastering the way various natural and artificial agents will interact among each other. Understanding extrinsic complexity typically leads to redesigning artificial agents (until a stable technology maturity is reached) and the definition of training human agents (until a stable maturity of practice is reached).

During the eighties, Airbus introduced glass cockpits, fly-by-wire and highly automated aircraft technology. Everything was done with a vision, but implemented incrementally using a strong positivist approach (i.e., functions, instruments and systems were added incrementally). This resulted in an accumulation of systems, and more importantly, their limited autonomy transformed the job of a pilot into a management job. Literally, the pilot's job moved from control to management. A new **phenomenon** emerged in the cockpit: systems management. Cognitive engineering was being developed at the same time and rapidly became the scientific support for the analysis of this kind of job evolution. In the making of complex systems it is then important that **systems engineering** (the positivist approach) be carried out taking into account **systems management** (the phenomenological approach).

Systems engineering (SE) focuses on developing processes, which need to be certified (e.g., ISO 9000 and CMMi).[7] SE approaches lead to form filling, database

---

investigated (Laudeman et al. 1998; Athènes et al. 2002; Masalonis et al. 2003; Hilburn 2004; Cummings and Tsonis 2006).

[7] CMMI (Capacity Maturity Model Integrated) was developed by the Software Engineering Institute at Carnegie Mellon University. This process improvement approach helps integrate traditionally

management, requirements analysis and traceability, model-based system design and performance modeling, planning and project control, software engineering and formal engineering documentation. Very sophisticated tools have been developed to support this kind of approach (e.g., CRADLE),[8] and leads to collaborative system engineering. However, this kind of engineering (positivist) approach cannot be successful without taking into account systems management (i.e., a human-centered design phenomenological approach).

Sometimes people believe that a systemic approach, such as quality assurance, to enterprise management guaranties product maturity at delivery time. This is not true. Reporting requirements could be satisfied while the reported job itself could not be performed correctly. **Syntax has become so important that semantics could play a second role**. In other words, since jobs are dichotomized and related through reporting, the content of the reports is sometimes not verified, the report suffices to be validated (except under pressure when money is at stake). For that matter, it is crucial that agents' expertise be analyzed, anticipated and allocated correctly. Such expertise may actually emerge from practice. Current systemic approaches to system engineering tend to automate the enterprise. It is therefore crucial to make sure that people involved in the use of this human and machine multi-agent automation have the right qualifications and are able to handle their task correctly. In particular, it is important that **organization automation** (Boy and Grote 2011) clearly encapsulates authority sharing (function allocation), worker motivation, and is clearly understood and accepted cooperative work. Goals should be shared and commonly accepted. Based on these premises, very simple tools are typically sufficient to support group activities. People are usually not fully involved in jobs where they do not understand why they need to do things they are asked to do (i.e., where system complexity hides the overall goal and the need for their participation).

In order to re-establish semantics in design and development, it is important to discover the phenomena involved in the systems being developed. Why do Apple products work, and why are they sold all over the world? This is because they are easy to use, they are reliable and usually mature; they create very enjoyable user experiences, they are aesthetically pleasing, and so on. The phenomena behind these tools are typically related to safety, efficiency and comfort. Their intrinsic complexity is certainly very high, but their extrinsic complexity is low (i.e., anybody can use them). More importantly, people do not hesitate to buy them even if they are more expensive. Another example is Google where searching for information is so simple that we now use the verb "google" (e.g., "I will google it to find out more about it"). Google software is certainly very complex (intrinsic complexity), but its use is very simple, efficient and comfortable (extrinsic complexity).

But what caused the success of Airbus, Apple and Google? People! These are people who manage and, to some extent, own these companies. Airbus was created by

---

separate organizational functions, set process improvement goals and priorities, provide guidance for quality processes, and provide a point of reference for appraising current processes (www.sei.cmu.edu/cmmi).

[8] http://www.threesl.com.

a handful of competent and motivated people. They did not own Airbus per se, but they were the real leaders of it. They invented Airbus and made it a success. They worked on a vision (i.e., phenomena such as flight management, and not on short-term money-centered predictions). Apple's co-founder Steve Jobs was a charismatic leader who knew about extrinsic complexity and delegated intrinsic complexity to exceptional experts with a coordinated vision. Google is a new venture lead by real experts in the field of search, and who have exceptional skill in discovering how performance can be improved through motivation and involvement. Google's owners also understood that an information search must be free of charge! These are phenomena. They understood that systems engineering is second to systems management.

If it is important to say that visionary people are crucial, it is also important to investigate the kinds of tools and processes that can support such phenomenological approaches. But before introducing tools and processes, let us describe two useful concepts in more details: affordance and emergence.

It is interesting to observe how new technology leads to a variety of new activities that may often become persistent. Sitting in an airport, I watched three ladies in front of me texting on their cell phones; they never stopped until we were called to embark. One of them was even simultaneously texting on two cell phones. People need to communicate. Since this media **affords** one to do it, it is more than normal that this media is used. In addition, it is boring to wait for a flight in an airport, and texting seems to provide a great opportunity to bridge the gap and ensure a continuity with the normal way of interacting with people while isolated in such circumstances. It was also interesting to notice that even when these ladies put their cell phones in their bags, almost systematically the cell phones called them back, ensuring a continuous interaction.

In addition to the affordance concept, this very quick ethnographical account of cell phone use brings to the forefront another concept, **emergence**. Cell phones afford texting because it is easy to do, cheap and silent (does not disturb neighbors as voice would). People text because it is affordable. In addition, cell phone use is a source of emergent behaviors. Before cell phones, we were not texting! These types of **phenomena** (e.g., texting) emerge from the use of new technology (e.g., cell phones). Whenever such emergent practices become predominant and stable, we generally talk about maturity of practice together with a mature technology. Such phenomena cannot be anticipated (predicted). This is because they emerge from chance and necessity (Monod 1971; Atlan 1970). Emergence of texting as a persistent phenomenon is based on four very simple facts: cell phones are equipped with texting capabilities; cell phones are easy to use for texting; texting is cheap; and texting is silent. We are able to rationalize these three facts afterwards, but did someone think about this combination of facts to make texting a success? I guess nobody did. Texting as a phenomenon emerged from the complexity of our evolving interactive technological society.

It is amazing to observe that modern cell phones have tremendous computing power (i.e., are made of incrementally-engineered software and hardware parts in the positivist sense, and their use induces various kinds of persistent phenomena, texting for example). Another example of emergent phenomena in the use of cell phones is the disappearance of planning in our everyday life. People rely on cell connectivity instead of planning (e.g., instead of using a shopping list, they call their spouse to ask in real-time what is needed from the grocery store). GPS installed in cars tend to remove the natural ways of searching our way on the road. We rely on GPS because it is a mature instrument that leads to mature practices. However, issues arise when this type of system fails. We can experience such **dependency** when they fail at supporting, guiding and/or mediating us; we can be literally lost.

Human-centered design needs to take into account these phenomena. Since they are emergent by nature, **human-in-the-loop simulation** is a necessary tool and method to discover these phenomena prior to developing the system being designed. For a long time, aeronautics experimental test pilots were the only resources to assess and provide the right modification in aircraft design from this point of view. Today, modeling and simulation tools and methods provide capabilities prior to any product development. The main issue that remains is the use of these tools and methods by potential real end-users. In the same way, I watched the three ladies texting, we should be able to play such scenarios using current modeling and simulation tools and methods.

Modeling and simulation are now inevitable in human-centered design. Until now, ergonomists and human factors specialists were used to evaluating system usability just before delivery. . . and sometimes after to explain incidents and accidents. During the nineties, usability engineering (Nielsen 1993) was extensively developed and commonly used in design. The concepts of horizontal and vertical prototypes were used to test systems globally in breadth and locally in depth. Today, we have modeling and simulation tools that enable us to test usability, in the sense of safety, efficiency and comfort, and during early stages of the design process. For example, the Flacon 7X was fully modeled and simulated before anything was built, and Dassault used the results of such modeling and simulation efforts in the development of the real aircraft. Modeling and simulation are useful to uncover the phenomenological patterns that will actually shape both structure and functions of the system to be developed.

The Difficult Issue of Human-System Integration

We have seen that complexity in complex systems usually comes from an un-controlled accumulation of systems that may be individually justified but may tremendously increase operational complexity as a whole. This kind of situation typically results from a lack of appropriate **human system integration** (HSI). In this case, HSI is about articulating system management. In highly interconnected systems where people and technology share authority, HSI needs to be considered from the early stages of the design process and throughout the product life cycle. When we talk about a product we also focus on its organizational environment. Again, design is about **technology, organizations and people** (the TOP model).

The outside-in approach promotes the necessity of having **architects** to ensure **integration** from the beginning of the design process. This approach does not dis-qualify builders who are important assets in the physical human-system integration to ensure details and robustness of the parts. Architects know how to define integration-driven requirements for builders. Architects typically have a scenario-based approach that integrates technology, organizations and people. They need to master both creativity and rationalization (i.e., they are both artists and engineers). In other words, they need to have a vision inspired by the scenarios and requirements from their potential clients, and technical skills that guarantee concrete construction of their mock-ups. In designing complex systems, we need architects who are curious and able to capture what is really needed from a TOP point of view.

Mathematicians acknowledge abstract objects such as triangles or lines. They manipulate these objects. They combine them to make more complex mathematical objects. They have a notion of "space". A Sobolev space (named after the Russian Mathematician Sergei Sobolev), for example, is a vector space of functions equipped with a norm that measures both the size and smoothness of a function. Partial differential equations are naturally found in Sobolev spaces. Another example is the mathematical concept of a Hilbert space (named after the Russian Mathematician David Hilbert) that generalizes the notion of Euclidian space with any finite or infinite number of dimensions. It can be used to study the harmonics of vibrating strings for instance. This notion of space is very useful to simplify the problem space. In a similar way, human-centered design needs to have this kind of problem resolution space.

Industry traditionally used the army type model as a problem solving space. A general is at the top of a cascade of officers down to soldiers. In this kind of space, no transversal communication is theoretically possible. Information flows are generally downstream, with a few information flows going upstream. Industry has evolved for many reasons that include service-oriented production, management of objectives and over-specialization of work. Instead of soldiers, we now have specialized musicians. The Orchestra Model is progressively replacing the army model (Boy 1991, 2009); this evolution will be more explained in Chap. 6. Let us describe the Orchestra Model as a problem solving space. In order to play a symphony, musicians need to understand the same music theory, read scores that are written by a composer, be coordinated by a conductor, and be proficient in one instrument. These four norms are mandatory. Using this analogy, we can rapidly envision the trouble of some of our contemporary companies. Do they have a music theory? Do they have composers? Do they have conductors? Are the musicians trained to play with others? These are many questions that require clear answers, taking into account that the old army model often overlaps the new orchestra model.

Three Examples of Complex Systems

This chapter cannot be terminated without taking three examples of complex systems that our twenty-first century currently develops: genetically modified organisms (GMOs), terrorism and finance-driven corporations. How did we end up making such complex systems?

GMOs are products of science, a combination of biology and computer science. We can make plants that resist weather (e.g., protect plants from frost damage, and other kinds of external aggressions by altering their DNA molecules). New genes can be created. Natural things are now artificial! We ended up creating complex systems that improve plant protection, and agriculture benefits.

If this kind of technology is of interest because agricultural production can be controlled more easily, economical control of such production is now strongly dependent on highly profitable companies that have the power to directly control local agriculture worldwide. When a country accepts this technology it becomes dependent on it for a long time (extrinsic complexity). We do not know the impact of such products on human beings medically speaking (intrinsic complexity). Product maturity is probably not yet reached, and science needs to tell us. For users, maturity of practice requires more attention since some countries have accepted GMOs and others are still reluctant.

The twentieth century was the century of two conventional world wars, plus a cold war, with clear divisions between the war makers. War makers were well identified with clear ideologies and objectives. Since 9–11, norms have changed. We passed from a war space to another one. War makers are more difficult to identify. They are likely to act on civil grounds with no anticipation. Uncertainty is more extreme compared to years past. In addition, we sometimes do not know if it is military or civilian activities. Complexity comes from ignorance, and opponents count on it. Complexity is also generated by solutions that the occidental world has set up to resist terrorism. Terrorist plans are almost always new. They act independently from each other. It is difficult to find a central organization behind actual actors.

We sometimes tend to design defenses that are obsolete by the time of use. The Maginot Line was built after World War I to protect France from possible German invasion. This defense assumed that war would be static and defensive like World War I. Unfortunately, the Germans invaded Belgium and consequently went around the Maginot Line, which became totally ineffective for this type of dynamic and offensive war.

Terrorism is not new. So why has it become a modern fact? A new type of complexity emerges from the way information is conveyed by media. Everything is going faster, and fear is sometimes exploited to make information more "exciting". In addition, security control in airports has become a giant business. Airfare is more expensive, travel is more annoying, and security is not guaranteed. Complexity has tremendously increased in order to insure security, without satisfactory results. Why? It is the same problem as the fundamental problem of safety that we have treated in aviation. We are using technology and organization that we know, avoiding to investigate the nature of phenomena behind the symptoms. Why terrorism? Many answers can be given to this question. I strongly believe that people need tangible spiritual models that give them a reason to live. When this is not the case, people are tempted to join groups that show strong beliefs, whether they are ideologies or religions, for example. It is therefore easy to train them and make them terrorists. The solution is not in the short-term as most of us believe. It is in the longer term by re-creating harmony, empathy and cooperation among people. What is the new music theory that we need to set up? Who are the composers? Who are the conductors? What are the profiles of these new musicians?

Finance-driven corporations have progressively emerged from the evolution of the worldwide market economy. For the last 20 years, management in large companies drastically changed from technology-driven to finance-driven leadership. Whether it was in aeronautics, the car industry or other technology makers, leaders were almost always technicians, or former technicians who evolved into management. Business managers have taken the lead, and more importantly finance people run the show. Shareholders have the real leadership of corporations today, and systems engineering provides tools and techniques that literally automate organizations to achieve shareholders' goals. However, this automation is performed using almost only financial variables; technical variables have become peripheral and human requirements are almost not considered. Excel spreadsheets now constitute the mediating representation and tool through which employees interact with the overall system. It results in impersonal interaction among the various agents of the entire enterprise. Motivation tremendously decreased because work has been dichotomized to the extreme in order to enable appropriate top-down financial management. Music theory is only finance-based. Accountability is at the financial level and not at the technical level any longer. Reporting has also become more important than the work it reports. There is lack of technical composers and conductors, and musicians require motivation and leadership. Analyzing this evolution of our corporations worldwide as complex systems, we immediately observe that they very often have inside-out approaches to system design and development by lack of top-level socio-technical leadership. They are definitely not human-centered, nor they are developing human-centered products. The main issue is that people are "still there" to do some kind of work and use generated products.

It is time to react to this evolution and create frameworks that make human-centered design possible. We need to replace short-term prediction by **vision**! Short-term prediction works well with numbers locally, but does not work at all with our lives globally and in the long-term. We need visionary people who set goals (composers) and create the mandatory motivation that enables the construction of a better world led by leaders who respect people and nature (conductors). Human beings have always created and developed artifacts. It is now very important to focus on the complexity of our combined artificial and natural world.

References

Amalberti, R., & Wioland L. (1997). Human error in aviation. Invited speech, International aviation safety conference (Iasc-97) Rotterdam Airport, The Netherlands. In H. Soekkha (Ed.), *Aviation Safety* (pp. 91–108). Utrech: Vsp.

ASA (1994). *Au temps de Clément Ader*, ouvrage coordonné par l'Académie de l'Air et de l'Espace. ISBN 2-87717-044-6.

Ashby, W. R. (1947). Principles of the self-organizing dynamic system. *Journal of General Psychology, 37,* 125–128.

Athènes, S., Averty, P., Puechmorel, S., Delahaye, D., & Collet, C. (2002). ATC complexity and controller workload: Trying to bridge the gap. In J. Hansman, S. Chatty, & G. Boy (Eds.), *Proceedings of HCI-Aero'02*, Boston.

Atlan, H. (1970). *L'organisation biologique et la théorie de l'information*. Paris: Herman.

Bernard, C. (2000). Principe de médecine expérimentale (in French). http://www.laphilosophie.fr/ebook/Bernard,%20Claude%20-%20Principes%20de%20m%E9decine%20experimentale.pdf. Accessed 11 Jan 2012.

Boy, G. A. (1983). The MESSAGE system: A first step toward computer-supported analysis of human-machine interactions (in French). *Le Travail Humain Jounal, 46*(2).

Boy, G. A. (1991). Advanced interaction media as a component of everyday life for the Coming Generation. *Proceedings of the World Marketing Congress*. Tokyo: Japan Management Association.

Boy, G. A. (1998). *Cognitive function analysis*. USA: Greenwood/Ablex. ISBN 9781567503777.

Boy, G. A. (2005). Knowledge management for product maturity. *Proceedings of the International Conference on Knowledge Capture* (K-Cap'05). Banff, Canada. October. New York: Also in ACM Press Digital Library, (http://dl.acm.org).

Boy, G. A. (2009). The Orchestra: A conceptual model for function allocation and scenario-based engineering in multi-agent safety-critical systems. *Proceedings of the European Conference on Cognitive Ergonomics*. Finland: Otaniemi, Helsinki area, (30 September-2 October).

Boy, G. A. (2011). Cognitive function analysis in the design of human and machine multi-agent systems. In G. A. Boy (Ed.), *Handbook of human-machine interaction: A human-centered design approach*. Aldershot: Ashgate.

Boy, G. A., & Grote, G. (2009). Authority in increasingly complex human and machine collaborative systems: Application to the future air traffic management construction. *In the Proceedings of the 2009 International Ergonomics Association World Congress*, Beijing.

Bradshaw, J. (Ed.). (1997). *Software agents*. Cambridge: MIT.

Callon, M. (1991). Techno-economic networks and irreversibility. In J. Law (Eds.), *A sociology of monsters: Essays on power, technology and domination* (pp. 132–161) London: Routledge.

Cummings, M. L., & Tsonis, C. G. (2006). Partitioning complexity in air traffic management tasks. *International Journal of Aviation Psychology, 16*(3), 277–295.

Dooley, K. (2002). Organizational complexity. In M. Warner (Ed.), *International encyclopedia of business and management* (pp. 5013–5022) London: Thompson Learning.

Eisenhart, L. P. (1948). Enumeration of potentials for which one-particle Schrodinger equations are separable. *Physics Review, 74*, 87–89.

Ferber, J. (1999). *Multi-agent system: An introduction to distributed artificial intelligence*. Harlow: Addison Wesley Longman. ISBN 0-201-36048-9.

Ferber, J., Stratulat, T., & Tranier, J. (2009). Towards an integral approach of organizations in multi-agent systems: the MASQ approach. In V. Dignum (Ed.), *Multi-agent systems: Semantics and dynamics of organizational models*. IGI.

Ford, K. M., Glymour, C., & Hayes, P. J. (1997). Cognitive prostheses. *AI Magazine* (Vol. 18 Issue 3). Fall.

Grote, G. (2004). Uncertainty management at the core of system design. *Annual Reviews in Control, 28*(2), 267–274.

Hemingway, C. J. (1999). Toward a socio-cognitive theory of information systems: An analysis of key philosophical and conceptual issues, *IFIP WG 8.2 and 8.6 Joint Working Conference on Information Systems: Current Issues and Future Changes*. Finland: IFIP, pp. 275–286.

Hilburn, B. (2004). Cognitive complexity in air traffic control: A literature review. Project COCA—COmplexity and CApacity. EEC Note No. 04/04.

Holland, J. H. (1998). *Emergence: From chaos to order*. Reading: Perseus Books.

Hutchins, E. (1995). How a cockpit remembers its speeds. *Cognitive Science, 19*, 265–288.

Jamshidi, M. (2005). System-of-systems engineering—A definition. *IEEE SMC*, 10–12. http://ieeesmc2005.unm.edu/SoSE_Defn.htm. Accessed 20 Feb 2012.

Jordan, M. I., & Rosenbaum, D. A. (1989). Action. In M. I. Posner (Ed.), *Foundations of cognitive science*. Cambridge: The MIT Press.

Latour, B., (1987). *Science in action: How to follow scientists and engineers through society*. Cambridge: Harvard University Press.

Laudeman, I. V., Shelden, S. G., Branstrom, R., & Brasil, C. L. (1998). *Dynamic density. An air traffic management metric*. California: National Aeronautics and Space Administration, Ames Research Center, NASA/TM-1998-112226.

Lorenz, E. N. (1963). Deterministic nonperiodic flow. *Journal of Atmospheric Science, 20*(2), 130–141.

Luhmann, N. (1995) *Social systems* (trans: J. Bednarz & D. Baecker). Stanford: Stanford Press.

Luzeau, D., & Ruault, J. R. (Eds.). (2008). *Systems of systems*. Hoboken: Wiley. ISBN 978-1-84821-164-3.

Mandelbrot, B. (2004). *Fractals and chaos*. Berlin: Springer. ISBN 9780387201580.

Masalonis, A. J., Callaham, M. B., & Wanke, C. R. (2003). Dynamic density and complexity metrics for realtime traffic flow management. Presented at the ATM 2003 Conference, 5th EUROCONTROL/FAA ATM R&D Seminar, Budapest.

Minsky, M. (1985). *The society of mind*. New York: Simon and Schuster.

Mitchell, M. (2008). *Complexity: A guided tour*. New York: Oxford University Press.

Mitchell, M. (2009). *Complexity: A guided tour*. Oxford: Oxford University Press. ISBN 0195124413.

Monod, J. (1971). *Chance and necessity: An essay on the natural philosophy of modern biology* (trans: A. Wainhouse). Alfred A. Knopf (originally published as *Le hasard et la nécessité*. Paris: Le Seuil, 1970).

Nair, R., Tambe, M., & Marsella, S. (2003). Role allocation and reallocation in multiagent teams: Towards a practical analysis. AAMAS'03, July 14–18, 2003, Melbourne, Australia.

Nielsen, J. (1993). *Usability engineering*. Boston: Academic Press.

Norman, D. A. (1986). Cognitive engineering. In D. Norman, S. Draper, (Eds.), *User-centered system design* (pp. 31–61). Hillsdale: Lawrence Erlbaum Associate.

Norman, D. A. (2002). Complexity versus difficulty: Where should the intelligence be?, in *IUI'02 International Conference on Intelligent User Interfaces*. Miami.

Paulk, M., Curtis, B., Chrissis, M., & Weber, C. (1993). *Capability maturity model for software (Version 1.1)*. Technical Report CMU/SEI-93-TR-024.

Prigogine, I. (1997). *The End of Certainty*. The Free Press, New York.

Poincaré, J. H. (1890). Sur le problème des trois corps et les équations de la dynamique. Divergence des séries de M. Lindstedt. *Acta Mathematica, 13*, 1–270.

Rasmussen, J. (1986). *Information Processing and Human-Machine Interaction—An Approach to Cognitive Engineering*. Amsterdam: North Holland.

Reynolds, C. W. (1987). Flocks, herds and schools: A distributed behavioral model. *Computer Graphics, 21*(4), 25–34.

Schlindwein, S. L., & Ray, I. (2004). Human knowing and perceived complexity: Implications for systems practice. *E:CO, 6*, 27–32.

Sharples, M., Jeffery, N., du Boulay, J. B. H., Teather, D., Teather, B., & du Boulay, G. H. (2002). Socio-cognitive engineering: A methodology for the design of human-centered technology. *European Journal of Operational Research, 136*(2), 310–323.

Chapter 6
Organization Design and Management

Introduction

In this chapter, we will uncover the difficult problem of organizational influence on HCD, and potentially the influence of HCD on organizations. Many authors have already investigated and analyzed such problem. In the human error arena, Reason (1990) provided the famous Swiss cheese model. This model provides a macroscopic view of what it requires to prevent potential accidents by developing appropriate defenses. However, it does not tell us how the organization should be designed to manage socio-technical reliability. Kletz (2000) addressed the influence of senior management on functional risk, and its involvement in the detail of safety issues with close attention to output, costs, and even product quality.

Vaughan (1996) examined the origins of the 1986 Challenger space shuttle accident. She pointed out that the accident was mainly caused by social structures: "a mistake embedded in the banality of organizational life." Again, **organizational routine and complacency** are strong contributing factors leading to accidents. More specifically, Vaughan showed how "deviance in organizations is transformed into acceptable behavior." We are talking about the culture of the organization itself, and the role of the senior management that needs to ensure this safety culture glue among the actors involved. Of course, an accident does not happen from only one cause, but the main issues remain senior management's situation awareness and appropriate decisions.

We also need to be careful about the way we use statistics. NASA management estimated safety as 10^{-5} accident per space shuttle launch, and in contrast NASA engineers estimated safety as between 0.5×10^{-2} and 10^{-3} (Feynman 1989). In reality, it has been two fatal accidents on 135 launches (i.e., around 1.5×10^{-2} per launch). Of course in these cases, organizational decision-making was carried out as though management's estimate was correct. This brings to the front the issue of **management authority and competence**. Sometimes, the distance between top management and real technical competence can be large. This is why it is crucial to make sure that situation awareness and technical competence be present at the highest organizational level to insure enough "organizational situation awareness" (i.e., top

G. A. Boy, *Orchestrating Human-Centered Design*,
DOI 10.1007/978-1-4471-4339-0_6, © Springer-Verlag London 2013

Fig. 6.1 Organization genotypes and phenotypes of a product, and human-centered design feedback

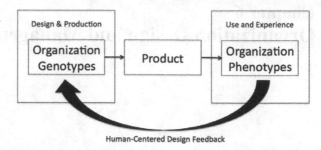

managers should clearly be aware of what is going on in their entire organization and more specifically at the product delivery level).

Human-centered design of a product cannot be developed properly if the organizational aspects are not taken into account in both the structure where design and development are performed, and the organization that will receive the product being designed. In the first case, we will talk about **organizational genotypes** of the product; in the second case, we will talk about **organizational phenotypes** of the product; ultimately HCD should take into account organizational phenotypes in order to influence organizational genotypes (Fig. 6.1).

Organizational genotypes include the way the organization influences product design, development, work assignment, articulation between technical and financial management, planning, quality control, evaluation, certification, external relations and personnel training (this is not an exhaustive list of course). Organizational phenotypes include the way the recipient organization integrates, operates and maintains the product.

System design and development is a matter of networked component production and integration. The organization producing the system can itself be defined as an **autopoietic system**, in Maturana and Varela's sense (Varela et al. 1974; Maturana and Varela 1987). An autopoietic system is a system, called System-1, that produces another system, System-2, that includes System-1's genes. In other words, an autopoietic system is producer and product at the same time. Following the autopoietic model, HCD is both the process of design and the design itself (i.e., a human-centered system design is incrementally modified in a participatory way by the various actors or agents who deal with it). HCD feedback is not only a matter of fixing the values of appropriate parameters; it contributes to regenerate the product itself. This biological definition of the genesis of a system encapsulates organizational genotypes and phenotypes. The HCD feedback takes into account emergent organization phenotypes in the redesign of organization genotypes and therefore the product itself.

In addition, computer networks irrigate current organizations, creating very complex connections among the various agents involved. Complexity of our organizations today comes from a mix of agents' diversity and number of connections among them. In this book, I would like to share with you my belief that current highly networked production systems work more like biological systems than what we use to think of as mechanically structured systems.

Reinventing "Educated Common Sense"

Autopoiesis is literally self-creation and discusses the relevance of the distinction between **structure and function**. Let us then explain the relationship between structure and function. For example, currently structural boundaries of organizations are getting fuzzier in our global economy; engineering functions are getting more dichotomized and standardized. Therefore, what are the associated organizational structures and functions that support better human-centered design?

As already emphasized in this book, when we design new technology, we design a trilogy of "Technology, Organization and People" (i.e., the TOP model), where "Organization" and "People" concepts mean that we also transform organization setups, as well as people's attitudes, practices, jobs and ultimately ways of living. Therefore, we understand HCD autopoeisis within the TOP model (i.e., Technology, Organization and People are the center of the constitutive dynamics of a living system, in Maturana's sense).

You may say that this self-producing machine analog does not work for industrial production systems since innovation tends to produce systems that are different from the original generator systems. This is a new facet of autopoeisis that deals with **mutation and evolution** in the Darwinian sense (Darwin 1859). Our technological society evolves (e.g., cell phones transformed the way we organize our lives and deal with other people), but we keep our basic constitutive societal dynamics almost intact, in the autopoietic sense. In other words, cell phones happen to be a chance and a necessity in our social evolution. In fact, there are many accidents that cause changes and consequently evolution. These accidents can be seen as chances and necessities (Monod 1970). The evolution of support of writing from clay, to papyrus, to paper, to printing, and finally to the Internet, is an example of such technological "accidents" that caused various evolutionary steps in our societies (Ganascia 2002).

We already know that living systems are determined by their structure. The **system-of-systems** (SoS) model can be used to approximate such structure. For example the World Wide Web is a SoS. In a SoS, components interact among each other to generate **emerging properties**. Component structures may change following these interactions, and consequently lead to more global changes at the level of the SoS itself. Sometimes such deformations may cause the generation of a SoS of a new kind; they also may cause splitting the original SoS into several new SoSs; finally they may cause merging the SoS with another SoS.

HCD has to take into account various transformations of our world induced by the rapidly evolving information-based society. These transformations are not simple changes, rather they are drastic ones. Information is now available to anyone, anywhere, anytime. Easy access is not only afforded to information repositories, but to other people also. We are in a global village. This may be overwhelming, and some of us may be lost in this informational hyperspace. This is why we need to reinvent new **"educated common sense"**. Common sense may have different meanings, going from judgment, to intuition, to ordinary language, to beliefs and wisdom. I prefer to add "educated" to "common sense" to reinforce inclusive knowledge and experience shared by a community of educated people. For example, it is "educated common sense" to predict what the weather will be tomorrow based on extensive accumulated experience and knowledge transferred from generations to generations in the form of mnemotechnical proverbs (e.g., "Red sky at night shepherds delight"). This was very well used by the ancient Greeks in their Art of Memory (Yates 1966). It seems that we are back to the Art of Memory with the Internet, mainly because related knowledge is based on incremental experience that is accumulated in the form of heuristics.

During the twentieth century, we built machines based on simplifications. Engineers have tried to simplify things in order to be able to develop technology, mainly mechanical technology. Simple systems can be linearized and therefore tractable techniques can be applied (e.g., linear algebra). We were using a **local/linear** approach. In this beginning of the twenty-first century, interconnectivity between people via technology has introduced new types of requirements that enable dealing with what Edgard Morin calls "complex thinking". The concept of complexity comes from the Latin term, *Complexus*, which means, "What is bound together". For example, a painter uses various kinds of colors and shapes, but each of these components is not meaningful at all before the whole painting is finished. It is the artistic glue that binds all components with each other that provides impressions, emotions and pleasure to the spectator. A clock and its parts are another global example. This holistic view of a product cannot be linearized. We need to use a **non-linear/global** approach.

Organizational Automation

In Chap. 2, organizational automation was introduced. Technological "automation" is a familiar term; autopilots, cruise control and the washing machine are examples of technological automation. The *"organizational automation"* concept reflects a progressively troubling set of issues. What are they? First, over the last decade or so, financial management has almost totally replaced technical leadership in large organizations; therefore, short-term profits have become more important than the global quality and sustainability of products and practice induced from the use of these products. Second, the never ending accumulation of layers of software in the systems that human operators have to handle in their work changes the way functions are allocated, to the point that we often do not know who is in charge or responsible. This is why the issue of authority sharing, distribution, delegation and trading has become increasingly requiring the development of a new kind of model for our socio-technical society to support the analysis, design and evaluation of such organizational automation. Third, we are losing technically competent employees who are being replaced by finance-based structures in which basic work is performed by low-cost, lower-skilled people; in the "automated organization" suppliers do the real technical work. Incrementally, technology integration has also become an issue. Fourth, people no longer want to take risks. They fear getting caught in a situation

Fig. 6.2 From army to orchestra

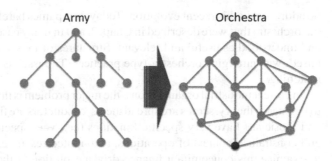

Army   Orchestra

where they might be breaking the automated-organization's rules. This is a legal and organizational cultural matter. In the face of uncertainty, it has become safer not to act, rather than risk making an error. Fifth, the development and integration of new technology and organizational changes introduce new types of complexity that are directly impacting perceived complexity and socio-cognitive stability. The many challenges associated with the automated organization's rules suggest new avenues for research on human and machine multi-agent systems. Hence, this book proposes the Orchestra organization model as support for such research endeavor.

Organizational Automation: From Army to Orchestra

A generic organization can be represented by a network of agents characterized by their cognitive functions (i.e., roles, contexts of operations and appropriate allocated resources). As already explained, such cognitive functions should be considered as societies of cognitive functions themselves.

Until recently, an industrial organization was typically managed as an army. CEOs were generals who sent orders to their lieutenants and so on until soldiers executed the basic tasks they were employed to execute. The typical organization structure was a tree of hierarchically organized cognitive functions. If such a hierarchy was vertically represented, information flows were always vertical and very seldom horizontal. Thomas Landauer describes this phenomenon as "organization of organizations" where the hierarchy of authority is flattened. He described the evolution of jobs done by "... highly valuable specialists who must be given adequate authority and access to the parts of the organization that depend on them. In turn, the increased flexibility and directness of the communication tools that information technology provides make obsolete the deep hierarchical command and control structures that originated in pre-electronic military organizations." (Landauer 1996, p. 122).

The evolution from army to orchestra is not straightforward as Fig. 6.2 may suggest. Even if in many cases emergent practices resembling orchestra-type practices have become very clear, most old structures persist. There are still many issues yet to be addressed in this recent evolution. Today, companies barely have the five pillars of the orchestra that were described in Chap. 2, and in many cases they are not perceived and understood as useful and relevant. Structures of most companies are army-type based, with emerging orchestra-type practices. The issue is then structure-practices matching.

In current industrial organizations, the major problem is that music theory turns out to be finance theory, scores are spreadsheets, conductors are finance-driven managers, and musicians have very specific functions (i.e., very specific roles, highly limited and constrained context of operations, and limited resources). This is why we need to examine this concomitant finance-driven evolution of the way organizations are managed.

The Current Power and Drawbacks of Financial Management

It is interesting to notice that technology is almost always ahead of science (not science fiction). Doing precedes rationalization. It takes visionary people to start doing great things. When John Fitzgerald Kennedy said in 1961 "I believe that this nation should commit itself to achieving the goal, before this decade is out, of landing a man on the moon and returning him safely to the earth", it was his visionary intention to involve a nation in the construction of a dream; this was tremendously risky and it worked! This is an excellent example that illustrates the very important distinction between a long-term human-centered vision and today's short-term finance-based developments. Even if this is not a new issue, it is important to note that today, even if technology is obviously based on good ideas, its development depends almost entirely on short-term funding possibilities and reactions to the market. I strongly believe that we now need more political motivations like Kennedy had in 1961.

Since then, automation and human-computer interaction have evolved to the point that our lives have become quite dependent on software. At the same time, technology enabled another type of emergence, the dominant financial management of companies. More usable computing power promoted finance. Finance has become so powerful during the last decade that it now manages whole enterprises, relocating engineering to a secondary role. It became clear that costs and financial benefits are more important than mature technology and its potential human benefits. However, this reality has started to show dramatic drawbacks by ignoring the overall question of co-adaptation of human and machine cognitive functions. We start to see delivery delays, maintenance issues and new training constraints.

Do technical-domain ignorance and "short-termism" characterize the state of our "modern" financial management? I propose to discuss and try to answer this very provocative question. The current evolution of finance-based enterprises worldwide justify that we take a careful look at our industrial and economical situation from an organizational automation viewpoint. The argumentation is based on my experience in aerospace so far.

Reporting and Financial Automation

The director of an industrial research company reported to me that he was asked by one of the shareholder representatives to "mechanize" his financial reports (i.e., define appropriate spreadsheet templates that provide a quick understanding of the financial state of the company to shareholders). This was in relation to organizational automation. He was required to provide a detailed report of research contracts and productions every month. He never understood that this requirement was so important until he understood that he was a part of a larger puzzle that needed to be financially articulated and coordinated. The main flaw in this approach was the fact that his shareholders barely asked him about the work his company was doing. Their only worries were finance-centered. They were looking for short-term financial forecast and potential benefits. Instead of defining and simulating possible futures (i.e., building a vision), they wanted to predict the future, which is almost impossible in research and innovation activities. Unfortunately, even if such short-term prediction often works, it leads to erratic organizational behavior, which sensible people, and mainly employees, often do not understand. In fact, finance-based administrators are not partial to long-term predictions because they were excessively based on unmanageable risks. Following up on this very typical example, it seems that visionary engineering leaders are needed to make long-term socio-technical decisions (i.e., we need decision-makers who are socio-technically competent).

The following is based on careful observation that industry managers are typically preoccupied by the reduction of costs and the increase in shareholders' benefits. This is true for manufacturers as well as for industrial customers that are required to operate and manage products in constant evolution. Since financial managers do not usually know about technical issues, they require appropriate reporting to keep their objectives within requested ranges. It is sometimes difficult to believe the degree of atomization of the spreadsheets that employees of these large organizations are required to fill in to justify their work. Work reports are detailed to the point that it is sometimes very difficult for employees to figure out what to put in the boxes. So what do these employees do? They do their best to enter what their supervisor is expecting, not necessarily what has really been done. Such reports may be very far from the reality that they are supposed to report. The worst result is that managers believe in these reports until it is often too late.

The rationale of reporting has deviated from its initial intent. Of course, it is important that a company knows about itself, and reporting is supposed to materialize this knowledge. Organizational memory is, however, not a static matter. It is highly dynamic. It should be constantly updated and verified. The main problem today is that this expert verification is essentially financial and is weaker in terms of technical expertise, where the quandary between reporting and actual work starts to emerge in an unexpected way. Reporting has become time consuming to the point that work may become secondary to work report itself. In addition, people know that what will be judged first and foremost is existence of the report, not necessarily its content nor, in the end, work efficiency. Consequently, we experience delays in the delivery of products, but more importantly decreasing work motivation among employees. Reporting naturally leads to the issue of storytelling.

Storytelling Versus Reality

It is commonly said that industrial power has moved from domain experts (e.g., engineers and experimental test pilots in the aeronautical industry), to managers. Furthermore, managers now need to get the right story at the right time in order to maintain a good information flow within the working structure among employees, customers and shareholders (i.e., three types of agents who do not necessarily know each other). **Reporting** becomes the full truth, and yet is always partial and decontextualized. Such a structuralist model works only until there are too many inconsistencies among the stories (i.e., between reporting and the reality). For example, the reality may be characterized by a clearly unacceptable delay in the production or product delivery. At this point, **reality** suddenly becomes the truth. The mismatch between the financial responsibility and the overall responsibility that includes technical responsibility is then revealed. We sometimes call this a scandal, but it is a systemic issue of the evolution of our managerial finance-centered society.

Storytelling is used by movie directors to make a film, and also by composers to write a symphony, in such a way that the final product is consistent with their intentions. Unfortunately, there are unexpected events that happen and may disturb the overall complex LCS. Handling complex systems is similar to living in an area where there are seismic activities and likelihood of earthquakes; nothing happens for a long-enough period of time, and people believe that it will never happen, but it does ultimately happen. This poker game is deliberate and the unfortunate end needs to be managed. When movie directors or composers of our industrial world are finance-based managers, they need to make decisions that may not be based on human and technical issues, rather on financial issues.

It would be much better for a community to write the overall story of a product as a blog, shared by the emerging community dealing with it, instead of focusing on its short-term financial attributes. This story would include design rationale, formative and summative evaluations, as well as experience feedback, for example. This kind of information would be very beneficial to all actors dealing with the product, and support human involvement. Only human experts could state design and evaluation rationale; this has a short-term cost but also has greater long-term benefits. Therefore, storytelling would become an excellent mediating support for communication, cooperation and coordination, as long as all actors know about its genesis, intentions, and limitations. It is a method for knowledge and strategy sharing. Gary Klein claimed that we ". . . organize the cognitive world—the world of ideas, concepts, objects, and relationships. We link these up into stories. By understanding how this happens, we can learn to make better use of the power of stories." (Klein 1998, p. 178).

What we miss in current industrial organizations is this domain-specific glue that puts people together, instead of isolating them into dry "quality" processes that, in practice, degrade the quality of communication among the parties. In addition, as already said, finance-driven reporting does not improve employees' motivation, which is the necessary fuel for communication, cooperation and coordination.

The Motivation Need

Among other things, human involvement has to do with motivation. Why? There are unpredictable situations that need to be managed, and this cannot be done without talent, competence, risk and enthusiasm. In addition, these surprises often create time pressure that needs to be managed. Again, storytelling and storyboards are a wonderful help for participatory design, where designers, manufacturers and end-users try to converge toward an acceptable and mature product. It is not surprising that many employees are no longer motivated and consequently involved in the work they are doing when they feel they represent a number in a spreadsheet box and they are told to apply the procedure without any possible interpretation. Of course, there is a wide selection of literature on motivation at work that would need to be investigated beyond the scope of this book (Maslow 1943; Festinger et al. 1956; Herzberg 1987), but I prefer to keep a simple approach to motivation such as appropriate feedback to employees (e.g., they need to see the effect of their work and be rewarded, and ultimately proud to belong to the organization they work for). This refers to the drive theory that claims people certain physiological needs and that a negative state of tension is created when these needs are not satisfied. It was inspired by the theories of Freud (1961) and system control theories.

In fact, motivation has a lot to do with educated common sense. This is what domain experts use when they are facing an abnormal or emergency situation. Reporting and obeying rules and procedures are constraining the use of educated common sense. However, when procedure does not fit the situation, educated common sense and critical thinking should prevail. By not regularly practicing critical thinking and using educated common sense, people are likely not to perform appropriately in abnormal and emergency situations, and crisis management. Procedures and rules cannot be written correctly for all possible contexts of use (i.e., human operators should be able to adapt to unanticipated situations). They obviously need to be motivated to succeed in this adaptation, and in many cases even start to implement this kind of adaptation. Consequently, there is a need for more expertise at work combined with the use of educated common sense.

Human involvement is a matter of availability for the care of the product being developed; an airplane for example, during its whole life cycle. This involvement should promote motivation for exploration, participation and confidence in a team. A single individual cannot do everything; good products are always the result of a team effort that requires encouragement and rewards. Excessive reporting and spreadsheet management tend to remove this necessary motivation because people do not see their involvement. Just like musicians who are able to hear the symphony they are all playing within the orchestra, employees should have this necessary feedback of their contribution to the company. In fact, people are more isolated into boxes related to quality processes as well as finance-based processes. They do not have to see what is outside their boxes! This is mostly because top management wants to insure integrity of the whole organizational automation.

This over-insured strategy is not only financial; it is also cultural. After September 11, 2001, security became a constant worry in airports. Not only airport taxes became outrageous, but also delays increased to the point that, for most of us, it is not fun to travel anymore. In fact, the overall airline transportation system's performance is directly impacted by the complex interaction of six elements (Donohue et al. 2008). These elements are airlines, security services, ATC system, government regulatory authorities, passengers and airports. Taking the Orchestra metaphor, these elements are musicians that try to play a symphony without appropriate music theory, neither conductor, nor composer. In fact, the only common frame of reference happens to be financial (i.e., music theory is finance-based). Safety and security are of course the major constraints.

In this overall system, nobody wants to take risk individually, unconsciously hoping that the system will stabilize by itself. Consequently, the overall system may block itself into a homeostatic working mode (i.e., it tries to preserve its own integrity until it eventually crashes). Such crashes may take various forms such as big delays, flight cancellations, heavy passenger frustration and/or discomfort, strikes and, sometimes, fatal accidents. "Unfortunately, these horror stories, and the many like them that occur almost daily, are not isolated incidents caused by random "bad luck" for which nobody is at fault. In short, we have reached a point of terminal chaos. . . " (Donohue et al. 2008, p. 2) Nobody is at fault because the chain of authority allocation is often a loop. In this overall complex system, there is no individual responsible person in charge of the overall operations and there is even less accountability to anyone else. Everyone is responsible for himself or herself, but there is no concept of responsible organization. In most cases, such an organization is politically correct since it often has a wonderful reporting system, syntactically irreproachable but semantically poor. Since such an organization is highly automated (i.e., there are appropriate means to calculate decision parameters), actors can react very quickly to events in the organization. Unfortunately, these decision parameters are almost always only financial.

Therefore, how can we get out from this catch twenty two problem where finance-driven management imposes its short-term cost-safety frame of reference to the whole industrial structure? The resulting organizational automation is now well in place and, in my recent experience, leaves very few degrees of freedom to employees. They are stocked into the system. Consequently, they seem to be less involved in personal actions but rather in codified acts. The short-term cost-safety frame of reference and the constrained function allocation remove any possibility of risk taking. How could a human being be involved if he or she is not allowed to take any risk?

Risk-Taking

Aeronautical research and industry never stopped improving machine reliability. They raised the aviation system to the very first rank of transportation modes on the planet, as far as safety is concerned.[1] A huge amount of research efforts have been carried out on human errors and the main result is that these errors are highly systemic (Funk et al. 2006). "In the systemic model, an accident can get prevented by understanding better how people and organizations normally function; how people and organizations do or do not adapt efficiently to cope with complexity" (Dekker 2006, p. 82).

Consequently, we need to focus on the human-machine relation, and more generally on the dynamics of the human-machine-environment triplet. Human beings have always searched to extend their hands and intelligence by making, adapting and using artifacts of all kinds; they never stop innovating and developing things to go further to improve safety, efficiency and comfort. To do this, they needed to take risks that they in turn needed to manage.

On February 2–4, 2008, my colleagues and I, from the Air and Space Academy in Europe, organized a conference on risk taking (ASA 2008). Six categories of issues were derived from expert discussions and debated during the conference with experts from various domains:

- (I1) cohesion and lack of cohesion of psychological and legal concepts of risk-taking;
- (I2) preparation of risky operations;
- (I3) risk and responsibility;
- (I4) individual and collective risk;
- (I5) risk and organization; and
- (I6) risk-taking in industry.

This understanding risk-taking has become increasingly important because one of us can and will have to take a risk and have to act one day! In this chapter, risk taking is encapsulated within the scope of organizational automation. We therefore need to better identify and understand the nature of constraints introduced by organizational automation. In order to progress on this, let us start to review a common misunderstanding in risk-taking, the zero-risk fallacy.

The Zero-Risk Fallacy

Bernard Ziegler, a test pilot and distinguished engineer, said, "routine is riskier than a first test flight" (ASA 2008). This statement was strongly based on a life-long

---

[1] However, we need to acknowledge that the ASRS, a public repository of aviation safety incident/situation reports from pilots, controllers and others, is still reporting that human errors remain the main cause of aviation accidents and incidents.

operational experience in the aeronautical field. Routine leads to complacency (Billings 1991). Due to the understanding and acceptance of this statement, the aviation industry comprehensively developed procedures and automation. Considerable progress has been made in aviation operations assistance during the last 4 decades. Conversely, extensive development of regulations has tended to remove human involvement, and increase the drive for compliance, including malicious adherence to procedures. As already said, we often loose sense making, judgment and more importantly detachment. Trying to decrease the residual parts of the certainty Gauss bell curve, we have pushed the unexpected very far, but we will never eliminate it! It seems we have forgotten that everything cannot be anticipated, and management of the unpredictable requires talent, competences, humility; without putting aside chance, exploration and openness to new things.

The time has come to override current logic of the search for zero risk that systematically blocks intuition and creativity. Too much focus has been placed upon individual-risk aversion that is likely to induce emergence of false fright, while collective risk-taking has increased tremendously in our societies and environment. Even this individual behavior is reasonably natural, organizational automation reinforce this behavior. A main consequence of this is that people do not move anymore because either the system is doing what they should do (i.e., over-assistance), or they are constrained to comply with rigid rules. Organizational automation is not a natural thing; it breaks because of it is made of rigid processes and tools that people are required to comply with and use. In this kind of approach, additional human work is always necessary to articulate processes among each other. When this articulation work is not done, the overall system becomes very brittle. For example, many appointments are typically taken by executive assistants who constantly need to articulate with their managers, who may in parallel take appointments at the time. It is quite interesting to notice that people who perform articulation work have to take risks, and they may not be recognized as they should be!

It is apparent that we are only interested in risk-taking in terms of profits and gains. This is a very cognitive approach to risk-taking, and more importantly an economical and financial one. We are facing a fundamental question; do we continue to focus on short-term gains while long-term risks are being disregarded? Current managers make extensive use of spreadsheets that include numbers representing short-term requirements and/or performances while the long-term vision is clearly forgotten. We are embedded into a very mechanized world of assurance. Constant worries are focused on regulation, reporting, calculations and so on. Despite this very strong influence of our socio-economical and socio-technical systems, human assets and instincts cannot be ignored. Even if the old adage "*errare humanum est*" is a long standing reality, it is remarkable to observe that error commission is an act of complete imperfection and must be totally eradicated at all costs. It is also remarkable that some of us have suddenly discovered that human beings actually do make errors and this must be taken into account as we emerge further into this human-driven mechanized society. Human beings have not changed fundamentally for millennia. What have drastically changed are the repercussions of their acts, whether erroneous or not.

Genuine progress, unfortunately, does not come without accident, error, failure or natural catastrophe. But it is urgent that we rehabilitate the right to make errors, to see the positive aspects of human error in a constructive way. Xavier Guilhou (ASA 2008) highlighted the fact that we should be able to pose real questions and address them the best we can, much like the Apollo 13 astronauts who said "Houston, we have a problem!" and did what it took to solve this problem, mutually, with their colleagues on the ground. In the same way, after hurricane Katrina in New Orleans, many of those who managed to survive were those who understood that the situation could be reversed and salvaged what they could with much human effort and their acts reflected this. This type of behavior was very similar to the mental and physical conditions that faced the Apollo 13 astronauts. Of course, some just had incredible good and bad luck, and results were exhibited by incredible human behavior. There are still other factors that need to be further analyzed such as ignorance, fright, confidence and boldness.

Motivation, Creativity and Leadership

Communication and information technology introduces new kinds of risk. Everyone is able to publish and prescribe using the Internet. Nothing is isolated anymore, and media influence increases virtual interrelations, which results in a noticeable rupture of people's confidence. Some individuals are frightened because everything goes fast, and we often forget that the stabilization of mature practices simply takes time. In addition, the media's job is to produce audience, not information. Television works on scoops that do not take into account the repercussions of the information that it delivers. It would like us to believe that our world can be saved in 45 min, but in reality regulations and bureaucracy slow down everything to the point that motivation decreases, and we have less tangible references (Guilhou, ASA 2008). For example, given that most of us spend about 8 h daily facing a computer screen, we barely have enough time to interact physically with real people. Consequently, our image of society reinforces individualism against group relations. We need experience feedback to further facilitate real interpersonal relationships.

Our society has a very difficult time admitting that we could be ignorant or rather ill informed. Xavier Guilhou (ASA 2008) said that our society favors defiance instead of **mutual trust**. Risk aversion battles against creativity, can kill projects, plus jeopardize success and progress. It is much more beneficial to accept the fact that our world is complex, that we may fail to notice things about it, and that we can trust each other to solve problems in it. Today, responsibility is excessively diluted, and therefore we are relentlessly busy trying to find an individualistic solution to an urgent situation. Where are responsible managers? We need to create the conditions to have involved competent leaders who are able to anticipate, interact and solve hard problems when they occur. This quest for human involvement does not remove necessary planning requirements and financial constraints.

Human Competence

In addition to physiological predisposition, it is inconceivable to imagine climbing Mount Everest without long-standing mountaineering experience and strong competence that is acquired when undertaking smaller endeavors. In order to take risks, you need to know what you are doing! Risk-taking depends on perception of the consequences of our actions. When we use automation, we need to appropriately know its role, its necessarily restricted context of use, and the appropriate resources necessary to fulfill its role (this obviously refers to the cognitive function concept). Missing knowledge of these attributes at any abstraction level can lead to problems. This is why the network of cognitive functions, which emerges from use of an automated eventually networked system, needs to be known by most actors involved in its use.

Conversely, nobody needs to know everything, but appropriate knowledge and know-how need to be accessible at the right time in the right understandable format, to the people who need them. There are cognitive functions statically allocated because of their persistent natures, and others that require dynamic allocation because what matters is context, and often context is very difficult to anticipate. In a complex organization, such a cognitive function network can be extremely dynamic and complex, requiring that the organization itself be dynamically structured accordingly. There should and must be various levels of coordination with appropriate competences at each node.

When a cognitive function becomes mature enough, it can be considered stationary stable in the environment where it is used. However, when it is not mature, its user must understand how the machine being used works. This is why, while automation is still not mature enough, it should typically be used or operated by experts who are able to understand it. For example, in the beginning of the twentieth century, many car drivers were also mechanics in that they could fix their simple machines. Today, most people do not need to know anything about the specifics of an engine or car body parts to drive a car. Consequently, there are machine cognitive functions, in the form of software, that become hidden from their human users (i.e., people do not need to know of these software functions), and we say that automation and more generally computers are invisible (Norman 1998). Today, major situation awareness problems arise when there are too many of these cognitive functions piled on top of each other.

Automation maturity is both a technology and a practice issue. From a technological viewpoint, engineers try to produce reliable, dependable and trusted automation. From a practice viewpoint, the use of automation involves situated knowledge and know-how (i.e., knowledge of the various kinds of situations in which its user needs to be informed, in order to do or not do things appropriately, when using the automation). Today, most of us want to do things right now, whatever the maturity of practice. This is inexact since it takes time to reach appropriate maturity of practice, especially using automated systems. The main question is about seriousness of the consequences and the ability to recover in a reasonable amount of time in order to maintain safety, performance and comfort. Product maturity and maturity of practice are incrementally reached through a co-adaptation process that may take time. We observe that in many cases today, such a co-adaptation process may end before one of these maturities (or both) is (are) reached. A new type of person emerges, the technological being.

Toward a Technological (Human) Being

We live in a technological society. Therefore, we cannot discuss automation and related human competences in the same way as before our current high-technology era. There are persistent human phenomena, often called human factors, which have never changed over the ages such as fear, empathy, stress, workload, emotion and anxiety. It is the nature of the triggering events and situations that have changed. Our environment is increasingly mediated by technology, to the point that we often need to make an effort to directly interact with someone face-to-face. Some of us interact with our neighbor colleagues, working in not so distant offices, using email or telephone. It would have taken only a moment to walk over and engage the individual personally! Is this practice good or bad? It depends on the situation. In aviation, there are situations where automation is used because it is better suited than manual control. However, it must not be used without knowledge of what it does in specific situations (Sarter and Woods 1995). Automation is expected to remove the burden of doing things that are tiring, repetitive and boring to do. Automation should be considered as a collaborator. Therefore, the management cognitive function required to properly interact with automation will need to be clearly understood, learned and properly used. In other words, automation does not remove the need for knowledge and know-how over ignorance. In particular, the context of validity of any useful cognitive function should be clearly understood and taken into account.

In the 1980s, new technology was developed for the upcoming glass cockpits (i.e., fly-by-wire aircraft), in a strong engineering environment. Technology was moving from mechanics and hydraulics to electronics and, more importantly, to computing. At the same time no one knew or understood that computers and software would invade cockpits so rapidly. In the mid eighties, the flight management computer (FMC) was introduced to support navigation. FMC shifted the role of pilot from control to management. He or she would have to manage computers (i.e., highly cognitive artificial agents), instead of acting directly on physical controls or skill-level automata such as the autopilot. Practice changed from control to management similar to the way someone gets a promotion and becomes a manager. He or she must learn new skills and competences to manage a team. However, when things go wrong, the pilot needs to be able to revert to manual control. Today, this cockpit computerization has become more reliable, dependable and trustworthy. What we better understand is the need for onboard system integration (i.e., these artificial agents that were managed "individually" are becoming more integrated toward a more naturalistic use). Obviously, the way this integration is done is crucial to providing good usability at operations time. In addition, redundant systems are available in case of failure of the primary ones. This does not remove the possibility of unexpected failure; this is why pilots are required to keep their flying capabilities.

Why Are These Changes Happening?

The washing machine, car power steering and elevators are examples of automation that were seen as progress. It was a step beyond industrial mechanization. This kind of automation is mainly based on basic control theories. It induced the shift from physical work to mental work, and induced the emergence of various kinds of applied cognitive sciences that are needed to support the human-centeredness of its development. Software progressively became the dominant resource supporting automation. Industry developed an increasing number of software-intensive tools such as text processors, spreadsheets, onboard systems, medical instruments, telecommunication applications and so on. The development of our worldwide economy is strongly based on the use of various kinds of automated services, and would not be sustainable without them.

New rules have been produced, based on the various possibilities that automation offers. Automation not only has enabled the ability to make routine jobs obsolete, but most importantly it enabled industry to "automate" relationships among workers. Office automation, for example, simplified several workflows. I remember the time when we needed to wait for our secretaries to type our hand-written research reports, when I started working at ONERA. No scientist or engineer would have imagined typing a report at that time. Text processing arrived, and it took a few years to make the typing job of secretaries obsolete. Now, everyone directly types on a laptop that is connected to wireless Internet. Text processors are equipped with spelling and grammar checkers, and more importantly revision mechanisms that enable sharing with others through the Internet. Computer-supported cooperative work (CSCW) is now a reality at work everyday. Jonathan Grudin and Steve Poltrock have developed and taught CSCW for many years, and showed the power of such technology for cooperation at work, coordination of activities, as well as in design and evaluation (Poltrock and Grudin 1994). Grudin (1989) introduced a paradox introduced by collaborative software in organizations, which says, "What may be in the managers' best interests may not be in the ordinary users' interests".

In parallel, industries have become larger and larger, giving tremendous power to major shareholders and top managers. In many ways, automation facilitates the management of such companies, and has therefore influenced their evolution: first the automation of production lines and today, automation of the management. However, its extensive use sometimes leads to stupid activities. For example, in a large company, a friend of mine reported to me that prior to the delivery of a very expensive product, he inspected this product and he found that two parts inside the product were not painted. He then went to the quality manager, who systematically checked his books and replied to him that according to his records they were painted. The quality manager considered that his job was done since the report was completed with a status (i.e., "painted"). End of story! Division of work is an old practice in industry, but this needs to be articulated. In the past, army-type organization was handling it. Today, with the coming but not yet fully installed, Orchestra-type organization, things have changed because technical conductors are rarely there. We have finance-driven managers instead, who barely know about the technical domain. For sure, something has to be done to really develop the orchestra-type organization with its five attributes (i.e., music theory, composer and scores, conductor, musicians' competence and of course the audience to assess performance).

Automation has always been related to safety, efficiency and comfort. As already said, most automata were developed to accomplish a task that had been performed by humans before, and in some cases, enabled people to do things that they could do without them. Typically, automation was developed because it was possible to do so, very rarely when it was necessary to do it. Today, this trend has changed. We have shifted from technological possibility to financial necessity. Indeed, the use of automation in industrial organization is now a strong necessity to support finance-driven organizations. Successful companies rapidly understood that both technological automation and organizational automation enabled them to decrease costs and increase profits. However, Landauer reported that "an aircraft instrument manufacturer installed a computer-based resource management system. The goals were clear, the economic analysis sound, the implementation thoroughly planned. Ten months later, there had been no efficiency gains. The problem? Use of the system was rigidly governed by management rules. Workers were reluctant to override the system's decisions even when manifestly wrong. Local shortcuts and workarounds remained secrets." (Landauer 1996, p. 330). Today, the approach has changed and work is outsourced in low-cost countries, since human work is more costly in occidental societies. Consequently, the overall management cannot be sustainable without technological and organizational automation of distributed activities worldwide. Again, articulation work is the key, and when it fails, the overall system crashes (i.e., there are delays, unfinished products or big losses of money, which is ironic when the overall goal was to make more money).

The broadly-recognized initial intentions to automate were based on safety assurance and production efficiency. Unfortunately, organizational automation rigidifies the overall system, as we have seen above. Therefore a strong discipline is required to manage highly automated systems when they are evolving in safety-critical environments. Indeed, automated systems do a marvelous job for what they are programmed for, but they fail at providing recovery strategies when used out of their contextual domain of validity. Consequently, when either inexperienced or insufficiently trained human operators need to assure such recovery strategies, the overall system fails dramatically. This is why, today, there is a tremendous need for technical expertise in the management of highly automated safety-critical systems.

But why has risk-taking become a necessity? We have seen that automation is assuring safety either by providing appropriate technology or operational rules to follow. We have also understood that such automation fails when it goes out of

Fig. 6.3 Working out of
cognitive function context of
validity

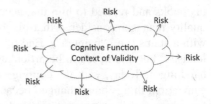

its definition context. Therefore, people who are in charge need to take risks, by definition, because they need to do something that is provided neither by technology nor by the required organizational rules. It requires motivated, educated, trained and responsible people to take risks. With respect to the definition of a cognitive function, risk-taking is when a human operator works out of his or her cognitive function context of validity (Fig. 6.3).

Getting out of cognitive function context of validity may mean that either the human operator is too tired or is not motivated enough or trained appropriately to accomplish his or her function (human side of the context), or the automation is not appropriate or is clumsy in the current situation of operations (technical or organizational automation side of the context). It could be compared to the notion of competence envelope (Woods and Hollnagel 2006). It is extremely difficult and sometimes impossible to predict everything, as far as cognitive function context of validity is concerned. It may even turn out that designers and/or manufacturers do not anticipate some cognitive function's context of validity that is induced by capabilities and limitations of technology, organizations and people.

Product Integration in a Large Organization

In June 2005, Airbus announced the first delay and notified airlines that deliveries would be delayed by 6 months. In reality, delays were effectively extended to 2 years since Singapore Airlines became the first airline in the world to commercially fly the double-decker A380 in October 2007. In 2006, as a consequence of this delay announcement, shares of EADS, Airbus's corporate parent, dropped 26 % and led to the departure of top executives. The delay also increased the earnings shortfall projected by Airbus through 2010 to 4.8 € billion.

Why this delay? The main cause was attributed to the concurrent design, configuration management and change control of a complex wiring of the cabin (100,000 wires, 40,300 connectors, and 530 km of wiring) (Heinen 2006; Kingsley-Jones 2006). Unfortunately, while French and British Airbus sites migrated to CATIA version 5 (the computer-aided design software typically used by most industries), German and Spanish sites kept using CATIA version 4, which caused major configuration management problems (Clark 2006). In addition, the use of aluminum instead of copper conductors required specific design rules that included non-standard dimensions. It happened that these management problems were not easily managed between versions of the CATIA software (Wong 2006). Why didn't top management anticipate this kind of incompatibility among work support tools? Engineering leadership is certainly at stake, but why?

A multi-national and multi-cultural company that produces large costly life-critical systems is a complex system. They are characterized by large numbers of actors, a variety of actors and intertwined connections among actors. Back to the Orchestra model, the CATIA version issue was a "music theory" problem as well as coordination of the "various scores supporting the symphony". This is an articulation issue related to a necessary **common frame of reference** (CFR). Again, a financial CFR is not enough, there should be a strong socio-technical CFR, in particular when we deal with life-critical systems. Who sets the CFR? The composer! In industry, we refer to engineering leadership.

What are good engineering leaders? First, engineering leaders must be recognized by their organization, no matter how large the organization is. A good team leader should be able to take criticisms and take them into account to modify both the organization and how it is managed. Recognition is typically based on domain competence, empathy, as well as decision-making capability and effectiveness. Socio-technical leadership involves effective communication and motivation maintenance among employees. It also involves conflict management, especially in intercultural groups. Function allocation is key in the sense that there should be a good match between professional people (functions), tasks and personalities.

Conclusion and Perspectives

At this point, one understands that we need to develop methods and tools to support human involvement, and culminates by dramatizing the fact that people make errors. We understand that creating work systems by trying to totally eliminate human error, we will not enhance motivation and promote expert human involvement at work. Excessive human error avoidance and technical work-arounds are not universal solutions when they decrease or remove human activity that is required to maintain mandatory situation awareness. The most important and powerful assets are expertise, competence and knowledge.

On January 15, 2009, flight US Airways 1549 was involved in an accident en route to Charlotte, NC from LaGuardia, NY. The pilot crash-landed the A320 airliner in the Hudson River. The crew was immediately hailed as five outstanding aviation professionals who performed in an exceptional way under extraordinary circumstances. The task was extremely constrained and did not leave any room for errors. This is the kind of situation in which leadership and audacious decision-making are crucial. The pilot and crew had to make decisions under pressure and maximum demand for quick thinking. Why did it work? The answer to this question is extremely difficult. There is a major factor involving both the expertise and experience of the pilot. Indeed, he was an expert glider, an experienced pilot (19,000 flight hours), and an aviation-safety expert. When we speak of knowledge and know-how assets, this is what we are talking about! The question of whether we should be thinking of training house-keeping personnel to fly an airliner is nonsense, even in our financially-depressed times.

We also understand that the issue of **authority** is central in the evolution of our human-automation environments in the beginning of the twenty-first century. Who is in charge and how? What are the chains of authority? How is authority controlled? It seems that the overall democratic system of authority needs to be revised in the face of information technology and society. We understand that people have become more specialized and, therefore, deeper knowledge is now mandatory to glue such people together, especially in our highly automated organizations. We need to better articulate "organizational automation" concepts because there are information channels and authority processes that either exist and are not clearly known by the actors involved, or simply do not exist. More research is needed to make these concepts and their interrelations explicit.

If we take the three interaction models that I already introduced (Boy 2002), sometimes we cannot go further than the supervisory model and need to acknowl-edge this. Often, we luckily develop mediation models. More rarely, we manage to communicate by common understanding. Why so rarely? A straight answer is because people are not involved; they are not proactive to place this kind of model into action.

There is no real life without **human involvement**. The entire history of our civ-ilization is based upon the spirit of initiative and risk-taking that have contributed to its evolution and progress. What would have happened if Clément Ader, Emile Dewoitine, Louis Blériot, Jean Mermoz or Chuck Yeager had not taken risks to do what they accomplished? This question could be easily generalized to anything that contributed to the comfort and progress of our everyday life. . . The personal history of each of us is grounded in risk-taking to some degree, that marks various stages of our existence and enables us to build, evolve and succeed in our lives. "Nothing ventured, noting gained" is an old proverb that is still valid! The legacy of risk-taking knowledge must be given back to scientists and creators, at the same time we need to rehabilitate educated common sense practice.

Of course, **educated common sense** practice must reflect an expert experience in the domain of this practice and not a naive behavior. For that matter, there may be conflicts created by deliberate authority assignment and authority emerging from expertise. Both control and accountability of an a priori assigned cognitive function (i.e., owned by someone or automation, may be questioned by another cognitive function emerging from adaptation in the operational context). We should not forget that individuals adapt to situations when initially assigned cognitive functions are either limited or inoperable. They may also discover, by themselves, new cognitive functions that may be good or bad. The real issue at this point is to rationalize and validate these **emerging cognitive functions**. It may happen that some of these functions might be dangerous and therefore must be forbidden and/or eliminated. In other cases, they might become part of new organizational regulations and technology requirements.

Alan Kay, the father of the laptop and windowing computer interfaces, said, "Don't worry about what anybody else is going to do... The best way to predict the future is to invent it. Really smart people, with reasonable funding, can do just about anything that doesn't violate too many of Newton's Laws!" Success is a good mix of knowledge, enthusiasm, confidence, courage, perseverance, and well-thought-out risk-taking. All this is human centered.

References

ASA (2008). *Academy of air and space conference on risk-taking electronic proceedings.* http://www.academie-air-espace.com/publi/detail.php?varID = 150&varCat = 1.

Billings, C. E. (1991). *Human-centered aircraft automation philosophy.* NASA Ames Research Center, Moffett Field, CA, USA.

Boy, G. A. (2002). Theories of human cognition: To better understand the co-adaptation of people and technology, in knowledge management, organizational intelligence and learning, and complexity. In L. Douglas Kiel (Ed.), *Encyclopedia of life support systems (EOLSS), developed under the auspices of the UNESCO.* Oxford: Eolss. http://www.eolss.net.

Boy, G. A. (2007). Human-centered development of perceived complexity principles: Developed criteria. Technical Report DGAC/EURISCO No. T-2007-201, France.

Boy, G. A., & Grote, G. (2009). Authority in increasingly complex human and machine collaborative systems: Application to the future air traffic management construction. *Proceedings of the 2009 International Ergonomics Association World Congress*, Beijing.

Clark, N. (2006). The airbus saga: Crossed wires and a multimillion-euro delay. *International Herald Tribune.* http://www.nytimes.com/2006/12/11/business/worldbusiness/11iht-airbus.3860198.html?_r = 2&pagewanted = 5. Accessed 6 Nov 2006.

Darwin, C. (1859). *On the origin of species* (1st ed.). London: Murray. ISBN 0-8014-1319-2. http://darwin-online.org.uk/content/frameset?itemID = F373&viewtype = text&pageseq = 1.

Dekker, S. (2006). *The field guide to understanding human error.* Aldershot: Ashgate.

Donohue, G. L., Shaver, R. D., & Edwards, E. (2008). *Terminal chaos: Why U.S. air travel is broken and how to fix it*, N. Allen, (Ed.). Reston: Library of Flight, AIAA.

Festinger, L., Riecken, H. W., & Schachter, S. (1956). *When prophecy fails.* Minneapolis: University of Minnesota Press.

Freud, S. (1961). *Civilization and its discontents* (trans: J. Strachey). New York: W. W. Norton.

Funk, K. H., Doolen, T., Nicolalde, J., Bauer, J. D.,Telasha, D., & Reeber, M. (2006). A methodology to identify systemic vulnerabilities to human error in the operating room. *Proceedings of the Human Factors and Ergonomics Society 50th Annual Meeting*, San Francisco, California, pp. 999–1003.

Ganascia, J. G. (Ed.). (2002). Back to the future knowledge management past and present (No. 196, issue 4). Diogenes: Blackwell.

Grande, M., & Visscher, M. B. (1967). *Claude Bernard and experimental medicine.*, Cambridge: Schenkman.

Grudin, J. (1989). Why groupware applications fail: Problems in design and evaluation. *Office: Technology and People, 4*(3), 245–264.

Heinen, M. (2006). *The A380 programme.* EADS. http://web.archive.org/web/20061103062416/http://www.eads.com/xml/content/OF00000000400004/0/74/41485740.pdf.

Herzberg, F. I. (1987). One more time: How do you motivate employees? *Harvard Business Review, 65*(5), 109–120.

Kingsley-Jones, M. (2006). The race to rewire the airbus A380. Flight International. July 18. http://www.flightglobal.com/news/articles/farnborough-first-news-the-race-to-rewire-the-airbus-207894/.

Klein, G. (1998). *Sources of power: How people make decisions*. Cambridge: MIT.

Landauer, T. K. (1996). *The trouble with computers: Usefulness, usability and productivity*. Cambridge: MIT.

Maslow, A. H. (1943). A theory of human motivation. *Psychological Review, 50*(4), 370–96.

Maturana, H. R., & Varela, F. J. (1980). *Autopoiesis and cognition; the organization of the living*. Boston: Reidel.

Maturana, H. R., & Varela, F. J. (1987). *The tree of knowledge: The biological roots of human understanding*. Boston: Shambhala.

Monod, J. (1970). *Le hasard et la nécessité. Essai sur la philosophie naturelle de la biologie moderne*. Seuil. ISBN 2020028-123.

Morin, E. (1990). *Introduction à la pensée complexe*. Paris: EST Éditeurs.

Morin, E. (1994). *La complexité humaine*. Paris: Flammarion.

Norman, D.A. (1998). *The Invisible Computer: Why Good Products Can Fail, the Personal Computer Is So Complex, and Information Appliances Are the Solution*. Cambridge, MA: MIT Press. ISBN0-262-14065-9.

Poltrock, S., & Grudin, J. (1994). *Computer-supported cooperative work and groupware. Tutorial, CHI'94 Conference Companion*. New York: ACM.

Reason, J. (1990). *Human error*. Cambridge: Cambridge University Press.

Kletz, T. (2000). An engineer's view of human error (3rd ed.). London: Institution of Chemical Engineers.

Sarter, N., & Woods, D. D. (1995). How in the world did we get into that mode? Mode error and awareness in supervisory control. *Human Factors, 37,* 5–19.

Sarter, N. B., Woods, D. D., & Billings, C. E. (1997). Automation surprises. In G. Salvendy (Ed.), *Handbook of human factors & ergonomics* (2nd ed.). New York: Wiley.

Varela, F. J., Maturana, H. R., & Uribe, R. (1974). Autopoiesis: The organization of living systems, its characterization and a model. *Biosystems, 5,* 187–196.

Vaughan, D. (1996). *The challenger launch decision*. Chicago: University of Chicago Press.

Wong, K. (2006). *What grounded the airbus A380?* http://www.cadalyst.com/cad/product-design/what-grounded-airbus-a380-10903 Cadalyst Manufacturing. December 6.

Woods, D. D., & Hollnagel, E. (2006). *Joint cognitive systems: Patterns in cognitive systems engineering*. Taylor and Francis. CRC, ISBN 0849339332.

Yates, F. A. (1966). *The art of memory*. Chicago: University of Chicago Press. ISBN 102 269 50018.

Chapter 7
Modeling and Simulation

Introduction

Designing a system requires both creativity and rationalization. These two processes are antagonist. The former is about divergent thinking (i.e., generation of ideas and brainstorming processes when it is done by a group of people). The later is about convergent thinking (i.e., analysis of ideas and synthesis into concepts, evaluation and prioritization of concepts). Generating concepts requires formalizing them in order to share them, and this is where modeling enters into play. We need to have the right conceptual tools in order to share ideas and concepts. This chapter introduces a few of these conceptual tools such as models.

What is a model? A **model** is a simplification of the reality (e.g., a system or an environment) we try to represent. Therefore a model is a simplified representation that puts forward a few salient elements and their relevant interconnections. If the model takes care of the interconnections part of the system, we need the **simulation** to take care of the interaction part. In other words, simulation brings the model to life.

Modeling and simulation (M&S), as a discipline, enables understanding of purposeful elements, their interconnections and interactions of a system under investigation. When developing a model, you are always facing a first tradeoff in choosing the right level of **granularity** (i.e., what is meaningful against what is unnecessary). If you stay at too high a level, you might miss interesting interactions during the simulation process. Conversely, if you want to model every detail, you might run into very complicated interactions extremely difficult to understand. The model of everything does not exist and will never exist. Stay focused on the **purpose**! There are various kinds of models, and we will see the main distinctions that need to be understood in order to use and/or combine them appropriately.

Simulation is typically computer-supported and generally interactive, but not necessarily so (e.g., it could be paper-based or part of a brainstorming as a role-playing game). As already said, simulation is used to improve the understanding of interactions among various elements of the model it implements and simulates. It is also used to improve the model itself and eventually modify it (Fig. 7.1). Simulators enable people to be engaged into the activity that the system being modeled enables them to

G. A. Boy, *Orchestrating Human-Centered Design*,
DOI 10.1007/978-1-4471-4339-0_7, © Springer-Verlag London 2013

Fig. 7.1 The M&S design cycle

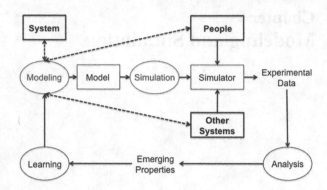

perform (e.g., driving a car simulator). Simulation can be useful for learning about a domain, such as flying, and simulators are typically upgraded as we learn more about this domain. In addition, simulation enables human operators to "experience" situations that cannot be experienced in the real world such as possible accidents. As shown in Fig. 7.1, modeling a system also takes into account people involved and interaction with other relevant systems interacting. The modeling process produces a model that can be run on a simulator, which in turn produces experimental data. These data are usually analyzed. Data analysis enables potential identification of emerging properties, which enables learning about system use. The M&S design cycle shows that modeling is a closed-loop process that in turn enables system re-design, modification of people practices/profiles and potential re-definition of the other systems involved.

Simulation is **imitation.** I remember the first time I discovered the Concorde simulator at Aeroformation in Toulouse, France. The view was generated from a small camera moving over a giant model landscape that was fixed to the wall in a big hangar. The images from this camera were projected onto large screens in front of the cockpit windows. The original simulation setup was limited to a single airport obviously! Then came computer-generated images that totally changed the possibilities of flight simulation. Note that this kind of simulator was only used for training.

Simulation can also be used to explain and aid difficult decision-making processes. For example, M&S was very much used at Kennedy Space Center to figure out launch pad configurations prior to a shuttle launch. Current M&S tools, such as CATIA and DELMIA for example, enable the visualization of complex structures and functions (Coze et al. 2009). NASA engineers and management use resulting simulations as mediating representations and decision support tools.

Should I mention Disney and Universal, for example, who use simulation for a totally different purpose (i.e., entertainment). In addition to attempting to provide natural sensations and "real-world" experience, they also create fiction. It is amazing how people can manage both simulated "natural" things and fictious objects brought to life through simulation. Computers and software make this kind of mix possible, but it requires creativity and experience to make this kind of technology acceptable and even invisible, to people and provide them with a memorable experience.

Fig. 7.2 The AUT triangle

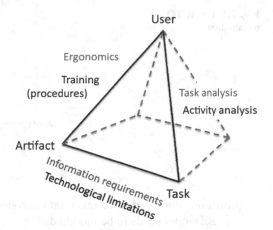

There are several kinds of simulation that involve both real and simulated people and systems interacting among each other. We typically use the term human-in-the-loop simulation in this case. Various kinds of effects can be simulated such as visual scenes, sounds, motion, and odors. The videogame industry is progressing very fast to integrate those sensations.

M&S principles, techniques and tools for human-centered design are introduced in this chapter. An HCD framework called the AUTOS pyramid will support them.

A HCD Framework: The AUTOS Pyramid

Whenever you want to design an artifact, you need to fully understand three entities: Artifact, User and Task. Artifacts may be systems, devices and parts for example. Users may be novices, experienced or experts, coming from and evolving in many cultures. They may be tired, stressed, making errors, old or young, as well as in very good shape. Tasks may be required at various levels including regular manipulation, process control, repairing, designing, supplying, or managing a team or an organization. Each task corresponds to one or several cognitive functions that users must learn and use. The AUT triangle (Fig. 7.2) presents three edges:

- Task and activity analyses (U-T): task[1] analysis is probably the first design initiative that enables a designer to grasp the kinds of things users will be able to do using the artifact; in contrast, activity analysis is only possible to implement when a first concrete prototype of the artifact is available.
- Information requirements and technological limitations (TA): task analysis typically results in a task model that contributes to the definition of information requirements for the artifact to be designed; however, there are always technological

---

[1] As a reminder, a task is a prescription and an activity is the effective realization of this prescription.

Fig. 7.3 The AUTO
tetrahedron

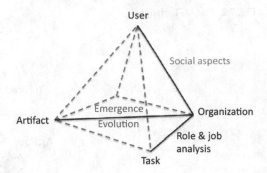

limitations that need to be taken into account—when they cannot be overcome, the task model needs to be modified.

- Ergonomics and training (procedures) (T-U): working on the ergonomics of an artifact means adapting this artifact to the user; conversely, defining training processes and operational procedures that contribute to adapting user to artifact.

In addition to the AUT triangle, one must also take care of the organizational environment in which an artifact will be used. The organizational environment includes all agents that interact with the user performing a task using the artifact (Fig. 7.3). Three more edges are then added to form the AUTO tetrahedron:

- Role and job analyses (T-O): introduction of a new artifact in the organization changes roles (i.e., cognitive functions) and consequently jobs, which need to be analyzed.
- Social issues (U-O): when changes induced by the introduction of a new artifact are drastic, social issues are likely to occur.
- Emergence and evolution of artifacts in the organizational environment (A-O): technology inevitably evolves and people get used to it, therefore a designer needs to take this evolution into account. At the same time, introduction of the new artifact creates emergence of new practices that need to be identified (i.e., emerging cognitive functions have to be discovered).

Of course, the AUTO tetrahedron needs to be tested in various situations. Consequently, the "Situation" is an important dimension taken into account, in the AUTOS pyramid (Fig. 7.4). Four more edges are then considered:

- Usability/usefulness (A-S): the artifact being designed and developed has to be tested in large variety of situations;
- Situation awareness (U-S): people need to be aware of what is going on when they are using the artifact;
- Situated actions (T-S): the task is not context-free and has to be consolidated in a broad set of relevant and purposeful situations;
- Cooperation/coordination (O-S): agents need to communicate, cooperate and coordinate within the organization according to a broad set of relevant and purposeful situations.

Fig. 7.4 The AUTOS pyramid

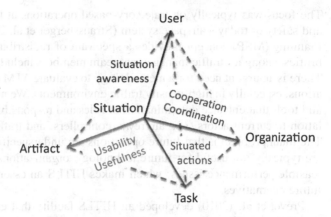

The AUTOS pyramid is proposed as a framework for categorization and development of HCD criteria and appropriate empirical methods. Each couple {criteria; methods} is associated with an edge of the AUTOS pyramid (e.g., the edge U-T is associated with task and activity analyses appropriate for analysis of perceived complexity of a system, device or user interface).

Human-in-the-Loop Simulation (HITLS)

By definition, HITLS requires human interaction with a simulator or a simulation facility. HITLS enables the investigation of situations that could not be investigated in the real world, such as incidents and accidents for example. When we think about HITLS, we immediately have in mind a flight simulator, but HITLS could be of any kind of tool including driving simulator, fashion design simulator, nuclear submarine simulator and so on. The development of HITLS requires methods and techniques for challenging systems and circumstances of the future (Boy 2011).

We will make a distinction between HITLS and simulated-human in the loop systems (SH-ITLS). The former directly involves real people in the simulation environment; the latter involves simulated people, avatars or human simulated agents. Of course, HITLS can a combination of both.

A primary purpose of HITLS is to effectively perform functions, roles and responsibilities allocation among various human and machine agents. Cognitive functions can be allocated to people or machines (typically software, as already described in the book). They are defined by their roles, contexts of validity, and resources that are cognitive or physical. For example, the PAUSA project (Authority sharing in airspace) goal was to gather insight into fundamental problems of human/automation integration and allocation of roles and responsibilities (Boy and Grote 2011) required to achieve significant capacity increases targeted for SESAR (Single European Sky Air Traffic Management Research program of the European Commission). This type of study is crucial to determine the future role of various agents in an airspace system.

The focus was typically on trajectory-based operations at triple the traffic density and safety of today's airspace system (Straussberger et al. 2008). The Multi-Sector Planning (MSP) concept provides a spectrum of redistributed roles and responsibilities among air traffic management team members including physical relocation. There is an urgent need for a methodology to evaluate ATM concepts and their variations, especially in high-density traffic environments. We need to develop methods and tools that enhance determination of roles and responsibilities of the MSP in relation to current actors (i.e., aircrews, controllers, and traffic management teams). The main problem is that future operations are underspecified. Prior best practices are typically not useful for future technology, organizations, and people to predict feasible performance issues, which makes HITLS an essential tool for evaluating future alternatives.

Prevot et al. (2010) developed an HITLS facility that enabled function allocation for ground-based automated separation assurance in NextGen. Other studies demonstrated the great utility of HITLS. For example, the FAA Free Flight Program successfully deployed the User Request Evaluation Tool (URET), Traffic Management Advisor (TMA), and Controller-Pilot Data Link Communications (CPDLC) to a limited number of Air Route Traffic Control Centers (ARTCCs) (Sollenberger et al. 2005). Enhanced HITLS capabilities can enable effective development of decision support tools (e.g., Traffic Management Advisor), procedural changes (e.g., Reduced Vertical Separation Minima), advanced concepts (e.g., Dynamic Resectorization), new software/hardware (e.g., Standard Terminal Automation Replacement System, Display System Replacement), and advanced technology (e.g., Global Positioning System).

HITLS enables understanding and development of relationships among technologies, organizations and people, and incremental refine configurations such as the Dynamic Airspace Configuration (DAC). DAC is an operational paradigm that represents a migration from static to dynamic airspace, capable of adapting to user demand and a variety of changing constraints (Kopardekar et al. 2007). DAC envisions future sectors to be substantially more dynamic and evolve fluidly with changes in traffic, weather, and resource demands. Traffic increase, mixed traffic (commercial, corporate, general aviation, unmanned aerial vehicles), and higher safety objectives require airspace to be globally and locally reorganized. The constantly evolving airspace system is one example of a complex multi-agent system, where emergence issues cannot be studied without a new generation of HITLS.

Simulated-human in the loop systems, such as MIDAS (Corker and Smith 1993), support exploration of computational representations of human-machine performance to aid designers of interactive systems by identifying and modeling human/automation interactions with flexible representations of human-machine functions. Designers can work with computational representations of the human and machine performance, rather than relying solely on hardware simulators with humans-in-the-loop, to discover problems and ask "what-if" questions about the projected mission, equipment, or environment. The advantages of this approach are reduced development time and costs, early identification of human performance limits, plus support for training system requirements and development. This is achieved by providing designers accurate information early in the design process so impact and cost of changes are minimal.

Multi-agent models and simulations are not new, even if many are still mostly created as ad hoc software. They have been used to study processes and phenomena such as communication, cooperation, coordination, negotiation, distributed problem-solving, robotics, organization setups, dependability and fault-tolerance. Such models enable exploration of alternatives needed in at least three major categories: (1) technology (to support such evolution and requirements), (2) organization (roles, context and resources, and multi-agent interaction models ranging from supervision to mediation and cooperation by mutual understanding), and (3) people (in terms of human capabilities, assets, and limitations).

Design Life Cycle

The design of a system is never finished even if at some point, delivery is required. This is why **maturity** has to be assessed (Boy 2011, p. 432). The more a system is being used, the more new cognitive functions emerge and need to be taken into account either in system redesign, organization redesign and/or training and operations support. Nevertheless focusing on the early design phase, and high-level requirements before development, four main processes are important to be described and taken into account: creativity; rationalization, information sharing and evaluation/validation.

Facilitating Creativity and Design Thinking

"The great pleasure and feeling in my right brain is more than my left brain can find the words to tell you." This statement of Nobel Laureate Roger Sperry illustrates the spirit of creativity. Innovative design is sometimes irrational and subjective, strongly based on intuition and emotion. This requires an open mind and holistic thinking. For that matter, design is an artistic activity.

Tim Brown defines "design thinking" as "a discipline that uses the designer's sensibility and methods to match people's needs with what is technologically feasible and what a viable business strategy can convert into customer value and market opportunity." (Brown 2008, p. 86). Brown, in his book "Change by Design", states that **design thinking** is human-centered innovation that uses people's needs in a technologically feasible and commercially viable way (Brown 2009).

Creativity involves taking risks since it is based on intuition, expressivity, least commitment and cultural blocks. According to Dreyfus (2002), **intuition** is a process that enables a person to do the appropriate thing without rationalization (i.e., without providing any explanation). In genetics, expressivity refers to variations of a phenotype in individuals carrying a particular genotype; in other words, creativity is expressed in terms of phenomena that are deeply rooted in the background of people generating this creativity. The "least commitment" attribute of creativity deals with the way ideas and concepts are generated; they can be described breath-first (a layered series of progressively refined shallow global descriptions) or depth-first (an assembled series of deeper local descriptions). Least commitment deals with both boldness and prudence. Creativity should tell you the right story at the right time. However, the analytical engineering culture may block storytelling. Constant searches for objectivity may block subjective thinking, and therefore design thinking.

Storyboarding usefully supports creativity, storytelling and design thinking. Storyboarding provides explicit visual thinking and enables one to share ideas concretely with others and refine them cooperatively. This is why cartoonists or similar should be part of design teams and support design thinking.

When such creative design is done by a group of people, it has to be orchestrated. The **Group Elicitation Method** (GEM) was designed to this end (Boy 1996a, b, 1997). Creativity was exploited in GEM by promoting innovation (the true original thinking), synthesis (combining information from different sources into a new pattern), extension (extrapolating an idea beyond current constraints and limitations), and duplication (improving or reusing an idea often in a new area or domain).

GEM is a brainwriting technique that can be computer-supported and enables contradictory elicitation of viewpoints from various (field) domain experts, augmented with a classification method that enables categorization of these viewpoints into structured concepts. Participants in a GEM session then score these concepts (i.e., each participant assigns a priority and an averaged consensus is computed for each concept). Ordered concepts are shared for a final debriefing where participants verify and discuss their judgments. A typical GEM session optimally involves seven experts (or end-users) and contributes to the generation of 10–30 concepts. Sometimes GEM is criticized because it provides very basic viewpoints from field people that cannot be abstracted into high-level concepts. It is important to note that an expert in the field who is familiar with abstraction-making facilitates the concept categorization phase. In addition, a structuring framework or model such as the AUTOS pyramid is very useful to further refine categorization of the produced concepts.

In the DGAC[2] research effort on novelty complexity in aircraft cockpits (Boy 2008), we carried out three GEM sessions involving 11 airline pilots, 7 cognitive engineers, and two structured interviews of two certification test pilots. Forty-nine raw concepts were generated. They were refined into a concept map (CMap) presented later on in the chapter. A CMap is not only a graphical presentation of interrelated concepts of a domain (Cañas et al. 2001), but also a very good, integrating framework for relating concepts among each other. In this approach to novelty complexity analysis, if a CMap was generated from an extensive user experience gathering effort, it is still an initial contribution that provides useful and meaningful directions for further research and development. We used it to derive high-level criteria for analysis of novelty complexity.

Alternative methods have been used such as the Blue Sky approach developed by IHMC (The Florida Institute for Human and Machine Cognition) that is based on

---

[2] DGAC is the French Civil Aviation Administration.

creativity-based workshops. For example on March 2–4, 2009, IHMC ran a Blue Sky workshop for the NASA Exploration Systems Mission Directorate (ESMD) at Johnson Space Center (JSC), as one of the many meetings of that kind devoted to development of operational concepts and designs of the Lunar Electric Rover (LER), now renamed Space Exploration Vehicle (SEV). The goal of that meeting was to "visualize innovative LER displays and controls that handle information and activities in a connected manner." The main goal was to improve situation awareness and control of the LER by an astronaut. Based on experience of driving the LER and more specifically watching an experienced astronaut driving it, I proposed the "virtual camera" concept during that Blue Sky workshop, which was very much discussed and subsequently developed (Boy et al. 2010). It took a while before the virtual camera became a crisp and acceptable concept; it took many graphical description, papers and simulations to provide main attributes and relevance for its actual implementation. The virtual camera concept is described in Chap. 8.

Modeling for Rationalization

Design thinking has also to be rationalized in order to be sustainable. The conceptual model of a system being designed needs then to be rational and objective, strongly based on facts and argumentation. This requires designers to be structured and logical. For that matter, design is a scientific activity. Consequently, modeling and simulation take on a different face.

Discrete event methods are used to model and simulate chronologies and sequences of events (Robinson 2004). Modeling approaches go from finite-sate machines, Markov chains in particular, to stochastic processes and queuing theory. They use computational techniques such as computer simulation, the Monte Carlo method, variance reduction and pseudo random number generator. Discrete event methods have been used in industrial engineering and network simulation.

Multi-agent methods have been used for a few decades to computationally simulate interactions among agents. It is important to understand that, when the number of agents is very large, interaction between agents typically generates emergent behaviors at the macroscopic level. The main goal of multi-agent simulations is often to identify these emergent behaviors.[3] These methods use game theory, complex system approaches, complexity science, homeostasis, adaptation, autopoeisis and non-linear system theories (Myerson 1991; Prigogine 1997; Mitchell 2008; Cannon 1932; Maturana and Varela 1980). Multi-agent methods contribute to simulation of

---

[3] Note that fluid dynamics involve huge number of molecules, considered as agents. It has been modeled for a long time using partial differential equations that for example, take into account parabolic (diffusive behavior) and hyperbolic (convective behavior) factors. Simulation of these equations leads to discovery of emerging phenomena such as shock waves. A flock of birds is another example where very simple local mechanisms such as separation, alignment and cohesion enable the generation of emerging behaviors that look like what you can see when you watch a real flock of birds (http://www.red3d.com/cwr/boids).

complex phenomena and emergence of new behaviors. The chapter on complexity analysis provides useful inputs to better understand what these methods are based on.

When safety is at stake, reliability engineering is crucial. We always want to assess the risk (generally in terms of expectations of what can go wrong) and the consequences of a faulty life-critical system. Safety and dependability analyses are usually developed by taking into account failure scenarios, assessing probabilities of failure, ranking components according to their contribution to risk, and assessing magnitude of the consequences. There are several methods that can be used such as fault trees, states/events formalisms, Markov Graphs, and Petri Nets. These methods are very well-defined, easy to use, and supported by software as well as graphical representations. Issues arise when systems are very complex because these kinds of models become too complicated and difficult to maintain. Note that new solutions have been investigated, for example, Rauzy introduced a new states/events formalism, so-called Guarded Transition Systems that generalize both Block Diagrams and Petri Nets, and handle looped systems (Rauzy 2008).

Modeling for rationalization necessarily meets the "system approach that is both a way of thinking and a standardized international engineering practice, whose objective is to master industrial systems complexity in order to optimize their quality, their cost, their time to market and their performance" (Krob 2011). For Krob, the term "system" refers to both the industrial object realized through an industrial process and the highest point of view one can have when dealing with this industrial object. Associated with human-centered design, this system approach is in the broader sense sustainable.

Standardization is the ultimate challenge of engineering, usability and usefulness disciplines. More specifically, our fully distributed worldwide economy stresses the need for interoperability. However, what would be left to creativity if everything was standardized? Where will the local niches be that are so precious, and will determine the various cultures on this planet?

Modeling for Information Sharing

Good design requires information to be shared among various kinds of actors in both time and space, across various kinds of professions and corporate cultures.

Computer-supported cooperative work (**CSCW**) is now a discipline that enables the development of environments that support people in both their individual and cooperative work (Grudin 1994). Collaborative M&S methods associate technology, organization and people (TOP) as other M&S methods with a greater emphasis on communication, cooperation and coordination. Examples are e-mail, instant messaging, application sharing (document interoperability in particular), videoconferencing, collaborative workspace, task and workflow management, blogging (that supports knowledge capture and management), and wiki sites (that progressively synthesize knowledge on various topics).

Traceability methods enable documentation of verifiable design history including chronological, multi-location and multi-actor links. When we want to master the life cycle of a product, traceability is an important technique; in particular, requirements traceability is key. Traceability typically includes cross-referencing, standardization of typical specialized documents, and restructuring capabilities in case of requirement changes. These methods should enable linking requirements to system elements, in both directions; capturing design rationale including domain content, responsibility and accountability; verifying consistency; requirements verification; and requirements change history.

Modeling for Evaluation and Validation

Workload is a model that supported a large set of evaluations in human factors and ergonomics. I began to learn and work with workload models during the early eighties when I was working on the two-crew-men cockpit certification together with DGAC and Airbus Industrie (Boy 1985). At ONERA, I developed the MESSAGE approach based on a multi-agent model of the aircrew, aircraft and ground systems. MESSAGE was a French acronym for Model of Crew and Aircraft Sub-Systems for the Management of the Equipments. The resulting model helped us to characterize various information densities that were useful in providing an interpretation in terms of workload and performance. Indeed, workload and performance models have been developed and used in experimental flight tests, and derive appropriate human factors measurements useful for certification purposes.

Situation awareness (SA) is a broad concept that tries to represent a high-level cognitive function. Despite a lot of work performed in the human factors field, especially by Endsley (1995a, b), SA still needs to be further defined and refined. Endsley derived the SAGAT method to assess situation awareness (Endsley 1988). This method is useful in domains where experience feedback is high. The relationship between competence and knowledge of human operators and the real situation at work is very important to understand. Expected SA is related to many things such as motivation, engagement, competence (skill set) and professionalism (airmanship), domain knowledge, regulatory rules, company culture, time pressure, physiological and cognitive conditions of human operators, unpredictability of environment changes and so on. Of course, there is always the "**Black Swan**" (Taleb 2007): who could have predicted that the autopilot of the 1972 accident of a Lockheed Tristar of Eastern Airlines that landed in the Everglades killing 101 of the passengers and crew (Flight Safety Foundation, Aviation Safety Network).[4] The aircrew focused on the nose gear issue without noticing a severe altitude problem. This tunneling problem

---

[4] Flight EA401 departed New York-JFK at 21:20 EST for a flight to Miami. The flight was uneventful until the approach to Miami. After selecting gear down, the nose-gear light didn't indicate 'down and locked'. Even after recycling the gear, the light still didn't illuminate. At 23:34 the crew called Miami Tower and were advised to climb to 2000 feet and hold. At 23:37 the captain instructed the second officer to enter the forward electronics bay, below the flight deck, to check visually the

is not common but may happen, and it is crucial that Technology and Organization must be designed for People (i.e., the TOP Model must be used). Safety is usually expressed in terms of probability, based on frequency of occurrence. We know that when some situations are not frequent at all, this kind of safety approach will fail. In terms of probability distribution, the Gauss curve typically applies. This is great when situations are very well known in advance and are included in the analysis and subsequent models, but when a situation never happened before, what should we do? The only possibility is to anticipate possible futures, and simulate them the best we can.

Decision-making is another cognitive function that deserves attention in cognitive engineering, human-centered design and more generally in our socio-technical world. Making a good decision is always the matter of stating a clear objective, framing an answerable question, using an analytic model whenever possible and otherwise using simulation, analysis at an appropriate moment, determining the appropriate level of complexity, stating the best assumptions in the model, exhaustive output analysis, and finally establishing a sustainable application. Decision-making can be supported by appropriate simulation methods such as discrete-event simulations (Thesen and Travis 1989).

More generally, modeling and simulation enables the elicitation of working processes that cannot be analyzed using paper and pencil, and brainstorming. Graphical interfaces enable designers to quickly figure out through M&S, the main problems that are at stake in the current design processes. Incremental re-modeling is a good practice that supports collaborative work and design convergence. In addition, simulation enables the incorporation of measurable indices that can guide the design and development process. These indices are also models.

Evaluating the use of a system can be done by various kinds of methods. Evaluation methods typically belong to two categories: objective and subjective methods.

---

alignment of the nose gear indices. Meanwhile, the flight crew continued their attempts to free the nose-gear position light lens from its retainer, without success. The second officer was directed to descend into the electronics bay again at 23:38 and the captain and first officer continued discussing the gear position light lens assembly and how it might have been reinserted incorrectly. At 23:40:38 a half-second C-chord sounded in the cockpit, indicating a $+/-250$ feet deviation from the selected altitude. None of the crewmembers commented on the warning and no action was taken. A little later the Eastern Airlines maintenance specialist, occupying the forward observer seat went into the electronics bay to assist the second officer with the operation of the nose wheel well light. At 23:41:40 Miami approach contacted the flight and granted the crew's request to turn around by clearing him for a left turn heading 180°. At 23:42:05 the first officer suddenly realized that the altitude had dropped. Just seven seconds afterwards, while in a left bank of 28°, the TriStar's no. 1 engine struck the ground, followed by the left main gear. The aircraft disintegrated, scattering wreckage over an area of flat marshland, covering a 1,600 feet × 300 feet area. Probable cause: "The failure of the fight crew to monitor the flight instruments during the final 4 min of flight, and to detect an unexpected descent soon enough to prevent impact with the ground. Preoccupation with a malfunction of the nose landing gear position indicating system distracted the crew's attention from the instruments and allowed the descent to go unnoticed." (Excerpt from the narrative included on the Flight Safety Foundation Website). http://aviation-safety.net/database/record.php?id = 19721229−0).

Both categories lead to the definition, development and use of models. Subjective evaluation methods can be based on subjective scales, for example the Cooper-Harper scale (Harper and Cooper 1986). The Cooper-Harper scale was initially developed to assess handling qualities of aircraft by experimental test pilots during flight tests. In this case, the model behind the evaluation method is a scale ranging from 1 to 10, each level being defined and subjectively assessed by the pilots or an external observer. NASA's TLX (Task Load Index) is another example of a subjective method for workload assessment (Hart and Staveland 1988; Hart 2006). NASA-TLX enables users to perform subjective workload assessments on operator(s) working with various human-machine systems. NASA-TLX is a multi-dimensional rating procedure that derives an overall workload score based on a weighted average of ratings on six subscales. These subscales include Mental Demands, Physical Demands, Temporal Demands, Own Performance, Effort and Frustration. It can be used to assess workload in various human-machine environments such as aircraft cockpits, command, control, and communication (C3) workstations; supervisory and process control environments; simulations and laboratory tests. Other similar methods are also used such as time-line analysis (Gunning and Manning 1980) based on a simple model of workload calculated by the ratio, available time t_A on required time t_R to execute a task ($WL = t_A/t_R$).

In contrast with subjective evaluation methods that are based on assessment performed by people, objective evaluation methods are based on physical measurements such as electroencephalograms, electrocardiograms, and eye gaze positions using eye-tracking technology. However, if these measurements are said to be objective, experimental results need to be interpreted to provide meaningful assessment in terms of workload, situation awareness or decision-making for example. Such interpretations are based on models. More specifically, eye tracking data can be interpreted using: a cognitive model which states that longer fixations of the visual scene on an object reflect subject's difficulty to understand that object (or the objects' features); or a design hypothesis: users will more easily understand color objects than gray ones (Stephane 2011).

Interaction, Complexity and Dynamics: Models and Theories

We saw earlier in this book that life-critical systems support our lives whether for safety, efficiency and/or comfort reasons. Now, if we want to model and simulate them, we need to know more about their main attributes. I choose to focus on three of these attributes: interaction, complexity, and dynamics. In this section, I emphasize how these attributes are already seriously grounded on various theories, models and representations . . . mostly in cognitive science, artificial intelligence, psychology, anthropology, philosophy and mathematics (Merleau-Ponty 1962; Dreyfus 1972; Gibson 1979; Maturana and Varela 1980; Winograd and Flores 1986; Suchman 1987; Varela et al. 1991; Hutchins 1995; Beer, to appear).

Interaction

There is always a degree of interaction with a given life-critical system. We interact with our car when we drive, and we interact with our clothes when we wear them; this is direct interaction. Direct interaction involves some kind of **embodiment**. The philosophy of mind emphasizes the role that the body plays in shaping the mind. However, there are life-critical systems that suffer from lack of such embodiment, or more precisely are remote and not accessible for most of us, and most importantly, cannot be controlled by us. Nuclear power plants and commercial aircraft belong to this category because they have a great impact on our lives, but we have no access to their control and management. We then need to trust the people and organizations that design them, build them, control them and eventually dismantle them. This delegation in the interaction can cause doubts when their design and use rationale are not clearly explained and understood. This is another case where modeling and simulation are important (i.e., providing information and knowledge in an accessible and participatory way).

Reason started a discussion on cognitive aids in process environments as prostheses or tools (Reason 1987). What is the level of embodiment of prostheses versus tools? It is interesting to notice that tools extend human capabilities (e.g., the hammer is an extension of our arm when we are hammering a nail, and we forget that it exists when we become proficient at hammering). The embodiment is perfect. Heidegger talked about *zuhanden* ("ready-to-hand") to characterize such a phenomenon (Heidegger 1962). However, when the handle breaks for example, we cannot use the hammer normally and we need to figure out how to use the broken tool to perform the task (i.e., hammering the nail); the broken hammer is not embodied anymore. The task becomes more cognitive. A problem has to be solved. Heidegger talked about *vorhanden* ("present-at-hand") to characterize such a phenomenon. Expertise is a matter of continuous adaptation of technology and people. Some of us are called "experts" because they have mastered specific tasks using appropriate specialized tools (usually technology, but they could be conceptual tools such as languages or methods and procedures). Hammers used by glaziers are not the same as hammers used by smiths! The symbiosis of people's skill sets and specialized technology is incrementally compiled and refined from interactions between people and technology.

Interaction usually happens between at least two entities (e.g., molecules interact among each other, people interact with their environments, and more generally agents interact among each other). This leads to the definition of an **agent** as an entity who/that interacts with its immediate environment, which is usually characterized by other agents. In addition, the concept of agent is recursive (i.e., an agent can be a society of agent in Minsky's sense (1985)). Agent's structure is therefore mutually inclusive, like Russian dolls. The agent model fits very well with the systems engineering approach of systems of systems (Luzeau and Ruault 2008). Beer describes an agent and its environment as coupled dynamical systems; the agent in turn is composed of coupled nervous system and body dynamical systems (Beer, to appear). Kauffman (1993) proposed a seven-level model of self-organization and selection in evolution that helps structure our living world. It starts with the chemical level where multiple molecules end up as enzymes, going to the biological level where multiple genes lead to cells, which themselves lead to organs (development level). The neurological level creates the emergence of concepts and maps from multiple modules. Then the psychological level generates specialists from multiple minds, up to the sociological level where multiple cultures lead to organizations, and finally at the ecological level where multiple species lead to niches.

The environment of an agent has properties that determine possible actions of this agent. Gibson (1977, 1979) coined the term "**affordance**" to denote the relationship between an agent (a human being) and its environment (i.e., some environment properties "suggest" some specific actions to the agent). The related field of study is called "ecological psychology". Object's affordances are defined as innate relationships (i.e., they are independent of agent's experience, knowledge, culture, or ability to perceive, and whether it exists or not). Doorknobs and door handles are usually common examples when we have to explain affordances. A horizontal door handle suggests pushing because it is located at the same level of both hands of the person reaching it and in the direction of the movement of that person; therefore their affordance action property is "push". A vertical handle on the side of a door suggests pulling because when the person grabs it, pulling is the logical movement following the gesture of the arm of this person; therefore its affordance action property is "pull". In this case, there are no conventions, rules or procedures; affordances are innate relationships between a person and an object. However in many cases, affordances can be learned (e.g., we learn how to stop at a red light). This is what Norman called "perceived affordances", which are related to object's utility and can be dependent on the experience, knowledge, or culture of the human agent (Norman 1988). The perception-action nature of affordances typically specifies the skill-based behavior described by Rasmussen (1986). Of course, when this interaction level does not work, usually because there is no embodiment, the agent has to learn and develop appropriate skills to better handle his/her/its environment in the future.

Complexity Modeling

We know that it is extremely difficult, and most of the time impossible, to predict evolution of the behavior of a complex system. What kinds of modeling and simulation means can we use to represent, analyze, understand and ultimately master a complex system? Formal models and theories of complexity have been developed over the last few decades. I recommend reading a good survey presented in the air traffic management domain (Xing and Manning 2005). In this chapter, I concentrate on usable models, theories, methods and techniques that enable modeling and simulation of life-critical systems.

Pylyshyn (1985) referred to the equivalence between **cognitive complexity** and **computational complexity**, and compared cognitive processes to generic algorithms. The choice of an algorithm is often made under contradictory requirements, such as understandability, transferability and modifiability of an algorithm. For example, Card introduced KLM (Keystroke-Level Model) (6) and GOMS (Goals, Operators, Methods and Selection rules) to study text processing in office automation (Card et al. 1983). KLM and GOMS enable the prediction of the required time to perform a specific task. They assume task linearity (i.e., tasks can be hierarchically decomposed into sequences). GOMS belongs to the class of analytical models, and works well in very closed worlds. Kieras and Polson (1985) developed the Cognitive Complexity Theory (CCT) as an evolution of GOMS. They proposed several measures of interaction complexity such as the number of necessary production rules and the learning time, as well as the number of items momentarily kept in the working memory in order to predict the probability of errors.

Rasmussen (1986) proposed the SRK model to capture three types of behavior (i.e., Skills, Rules and Knowledge). He also developed an **ecological approach** based on five levels of abstraction hierarchy. Vicente (1999) used this approach to develop the Cognitive Work Analysis (CWA) approach. CWA supports ecological interface design and emphasizes design for adaptation. Javaux and De Keyser (1997) defined cognitive complexity of a human-machine situation (in which specific tasks are performed) as the quantity of cognitive resources that a human operator must involve to make sure that tasks are executed with an acceptable level of performance. However, the quantity of cognitive resources is a very limited way to assess cognitive complexity without taking into account qualitative issues. Van Daele (1993) made another distinction between situation complexity and the complexity of task and operational goals. He relates complexity to constraints blocking task execution, remote character of goals, multiple goals to be satisfied at the same time, interdependent goals and environment attributes, multi-determination, uncertainty and risks.

Norman (1986) proposed a generic model that takes into account human actions, learning, usability and possibility of errors. He proposed the following concepts: **physical** versus **psychological variables**; physical versus mental states; goal as a mental state; and intention as a decision to act to reach a goal. He expressed interaction complexity in terms of execution gulf and evaluation gulf. In particular, the distinction between physical and psychological variables reveals complexity factors related to interaction induced by use of physical system and the task to be performed.

Amalberti (1996) analyzed complexity by making a distinction between **nominal** and **non-nominal situations**. He related interaction complexity to action reversibility and effect predictability, the dynamics of underlying processes, time pressure, the number of systems to be managed at the same time, resource management when the execution of a task requires several actors, artifacts representation, risk, factors coming from insertion of safety-critical systems in cooperative macro-systems and factors related to the human-machine interface, users' expertise and situation awareness.

In the MESSAGE approach (Boy and Tessier 1985), interaction complexity was assessed as **information-processing difficulty** in early glass cockpit developments. Several difficulty indices were developed including visibility, observability, accessibility, operability and monitorability. These indices were combined with tolerance functions, which were expressed as possibility distributions of relevant user-interface parameters. Subsequent work led to the development of interaction blocks to **model interaction chains between agents** in order to better analyze and understand the emerging interaction complexity of underlying operations (Boy 1998a). Interaction-block development requires the elicitation and specification of various interaction contexts, and therefore structuring various relevant situations. Five generic interaction-block structures were proposed including sequence, parallel blocks, loop, deviation, hierarchy and interaction-blocks leading to either weak or strong abnormal conditions (Boy 1998b).

Today, emphasis is on **systems of systems**, where the need for a multi-agent approach is mandatory in order to understand various phenomena such as emergence of new properties, cognitive and socio-cognitive attractors, various kinds of singularities and more generally non-linearities. If a complex system is represented as a society of agents interacting among each other, the prediction impossibility comes from agents' constant adaptation and the nonlinear interactions among them. Analyzing complexity can be done by visualizing concepts such as number of basic elements, components or agents, their variety, and the internal structure of the system.

Causality is crucial, but we need to make a distinction between direct causality related to linear systems, and systemic causality related to nonlinear systems (Batten 2009). Small causes may produce large effects. Consequently, we will be looking for modeling and simulation tools that enable the visualization of agents' status, and their relationships among each other. Complex systems have to be explored constantly in order to detect anomalies, symptoms, irregularities with expected behaviors and so on. The job of handling a complex life-critical system consists in observing appropriate cues, diagnosing and testing hypotheses. Epidemiologists developed methods to identify biological and behavioral causes of diseases, which isolate single causes for example. However, it is recognized that there are multiple causes and integrated models are often necessary.

People view complexity differently with respect to their background, domain competence and knowledge, and experience. For that matter, sharing and combining expertise is key. M&S tools that enable experts to share their **educated common sense** are good. Educated common sense can be represented by frames and metaphors that can be expressed in digital forms. Mastering the complexity of a life-critical system is often a matter of adjusting these frames and metaphors to the perceived reality, as well as sharing and combining them among experts. We already discussed the difficult question of "maturity". Reaching mature perception, understanding and mastering of the complexity of a system requires "nurturing", exactly like good parents nurture their children (Lakoff 2004). Nurturing has two aspects: empathy and responsibility. For that matter, M&S tools that enable a deep appreciation for another's situation and point of view will be good. Well-orchestrated participatory storyboarding, for example, is a good tool for the design of life-critical systems.

Dynamics and Control

Dynamic systems are systems that evolve in time. We are often mainly interested in their stability. Such systems have been extensively studied by control theorists and people who had to automate machines in order to improve safety, efficiency and comfort. The more we automate dynamic systems, the more we must make sure the resulting automated system is reliable, robust and dependable. We currently certify commercial aircraft systems with a probability of failure lower that 10^{-6} per h. Human reliability remains a main issue that we can take from two different perspectives: the negative one that says that people are the problem and we need to anticipate any human error because people will fail, no matter what, with a probability of failure greater that 10^{-2} per h; and the positive perspective that says people can be the solution since they not only correct most of their errors, but they are unique problem solvers and anticipators when they are very well trained and experienced. I qualify this second perspective, "human involvement and engagement." Of course, a mix of both perspectives prevails.

The concept of cognitive stability (Boy 2007) supports the concept of procedural interface (Boy 2002) that takes into account four main high-level requirements (i.e., simplicity of use as opposed to user-perceived complexity, observability /controllability, redundancy and cognitive support).

The interface of a system is characterized by a set of n observable states or outputs $\{O_1, O_2, \ldots O_n\}$, and a set of m controllable states or inputs $\{I_1, I_2, \ldots I_m\}$. The interface is redundant if there are p outputs ($p < n$), and q inputs ($q < m$) that are necessary and sufficient to use the system. The remaining $(n - p)$ outputs and $(m - q)$ inputs are redundant interface states when they are associated with independent subsystems of the overall system. These redundant states need to be chosen in order to assist the user in normal, abnormal and emergency situations. In aircraft cockpits, for example, several instruments are duplicated, one for the captain and another for the copilot. In addition, some observable states displayed on digital instruments are also available on redundant traditional instruments. Controlling a system state-by-state with the appropriate redundant information is quite different from delegating this control activity to an automaton. New kinds of redundancy emerge from the use of highly automated systems. Traditional system observability and controllability usually deal with the *What* system states. The supervision of highly-automated or software-rich systems requires redundant information on the "why", "how", "with what" and "when" in order to increase insight, confidence, and reliability: *Why* the system is doing what it does? *How* to obtain a system state with respect to an action using control devices? *With what* other display or device should the current input/output be associated?

Since life-critical systems are in general nonlinear systems, they will have several possible attractors (Ruelle 1989). We then talk about multi-stability. Sometimes these attractors are unknown in advance and strongly depend on initial conditions more generally on the history of the LCS. The brain as a complex dynamic system, whose structure does not change, can support many different attractors at the same time (e.g., concepts). Today, with the over-computerization of life-critical systems, cognitive engineers and human-centered designers should look for such attractors emerging from interactions in those human-machine systems they are designing and developing.

Design Evolution Versus Design Revolution

Is there a difference between designing from a blank slate and designing by modifying, and therefore improving, an existing system? The first distinction qualifies as design revolution, and the latter as design evolution. Improving the efficiency of a car engine is the effect of design evolution. Moving from horses to steam machines as transportation means, as well as moving from physical libraries and bookstores to the Internet, were design revolutions.

We need to be careful with the evolutionary course of modifications of a technology. For example, the incremental introduction of software in aircraft cockpits reached a point of revolution in the sense that increasing the number of artificial software-based agents led to a job change (i.e., moving from manual flight to systems management). It is always crucial to better understand what the emerging cognitive functions born from the interactions are with incrementally accumulated interactive software-based systems.

Therefore in any case, designers need to look for cognitive functions that emerge from the various interactions, whether it is manual control or system management. For that matter, modeling and simulation are required, human-in-the-loop simulation in particular.

Complexity and User Experience

Perceived complexity is about practice maturity. Product maturity analysis is also a matter of user experience with this technology. Is this technology appropriate for its required use? How and why, do or don't users accommodate to and appropriate the technology? Answers to this question contribute to a better understood perceived complexity, and further developing appropriate empirical criteria. Perceived complexity is a matter of the relation between users and technology. Interaction with an artifact is perceived as complex when the user cannot do or has difficulty doing what he or she decides to do with it. Note that users of a new artifact still have to deal with its reliability and availability, which are not only technological, but are also related to tasks, users, organizations and situations. This is why novelty complexity analysis requires a solid structuring into appropriate categories.

It would be a mistake to consider that a user who interacts with a complex system does not build some expertise. He or she cannot interact efficiently with such an artifact as a naïve user. For new complex tasks and artifacts, there is an adaptation period, full stop! Why? The answer is usually very simple. Complex artifacts such as airplanes or mobile phones providing personal digital assistants, cameras and other services, are prostheses. The use of such prostheses requires two major kinds of adaptation: mastering capacities that we did not have before using them (i.e., flying or interacting with anyone anywhere anytime); and measuring the possible outcomes of their use, mainly in terms of social issues. All these elements are very intertwined dealing with responsibility, control and risk/life management. Therefore, criteria that

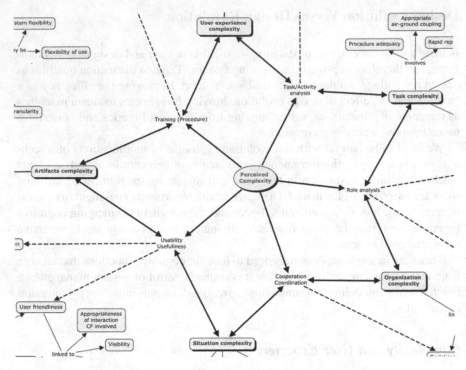

Fig. 7.5 Perceived complexity in the center of the AUTOS pyramid (CM-0)

will be given cannot be used for analysis by just anyone. They must be used by a team of human-centered designers who understand human adaptation.

Novelty Complexity in Aircraft Cockpits

Current literature does not address the issue of novelty that is naturally included in any design of interactive systems. We are constantly dealing with transient evolutions and sometimes revolutions without properly addressing the issue of coping with user-perceived complexity of new artifacts in interactive systems.

In this section, I use results of a study that was carried out in France under the oversight of DGAC (Boy 2008). The Group Elicitation Method (Boy 1996a, b) was used to identify the various ontological entities related to novelty complexity in aircraft cockpits. These ontological entities were further formulated into a concept map, or CMap (Cañas et al. 2001, 2005). The central concept of "novelty complexity" is connected to five sub-CMaps (Fig. 7.5). Relationships between the five first-level concepts characterize the edges of an AUTOS pyramid.

User experience (CM-1) concepts include training (expertise), trust, risk of confusion, lack of knowledge (ease of forgetting what to do), workload, adhesion and

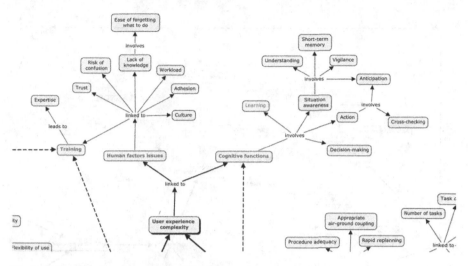

Fig. 7.6 User experience complexity (CM-1)

culture (Fig. 7.6). It induces several cognitive functions such as learning, situation awareness (that involves understanding, short-term memory and anticipation), decision-making and action (that involves anticipation and cross-checking). To summarize, a U-complexity analysis deals with user's knowledge, skills and expertise on the new artifact and its integration.

Artifact complexity is split into internal complexity and interface complexity. Internal complexity (CM-2) is related to the degree of explanation required for a user to understand what is going on when necessary (Fig. 7.7). Concepts related to artifact complexity are: flexibility (both system flexibility and flexibility of use); system maturity (before becoming mature, a system is an accumulation of functions—the "another function syndrome"—maturity is directly linked to function articulation and integration); automation (linked to the level of operational assistance, authority delegation and automation culture); and operational documentation. Technical documentation complexity is very interesting when tested because it is directly linked to the explanation of artifact complexity. The harder an artifact is to use, the more related technical documentation is required and therefore it has to provide appropriate explanation at the right time in the right format.

Interface complexity (CM-3) is characterized by content management, information density and ergonomics rules (Fig. 7.8). Content management is, in particular, linked to information relevance, alarm management, and display content management. Information density is linked to decluttering, information modality, diversity, and information-limited attractors (i.e., objects on the instrument or display that are poorly informative for execution of a task but nevertheless attract user's attention). The "PC screen do-it all syndrome" is a good indicator of information density (elicited improvement-factors were screen size and zooming). A clear and understandable language was the focus of ergonomics rules, error tolerance, redundancy and information

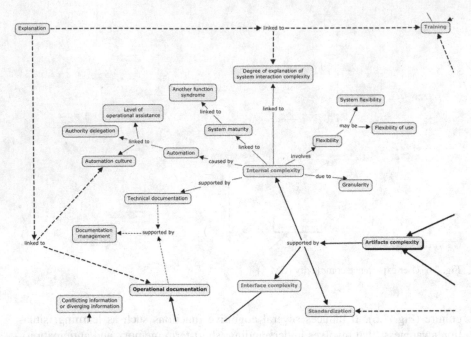

Fig. 7.7 Artifact complexity (CM-2)

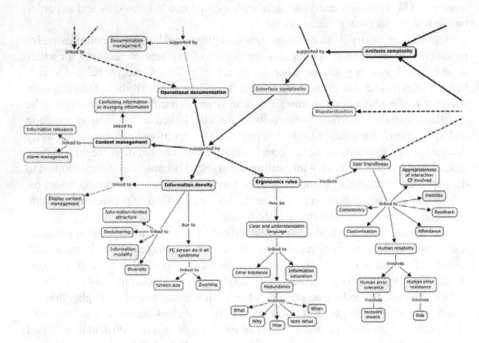

Fig. 7.8 Artifact complexity (CM-3)

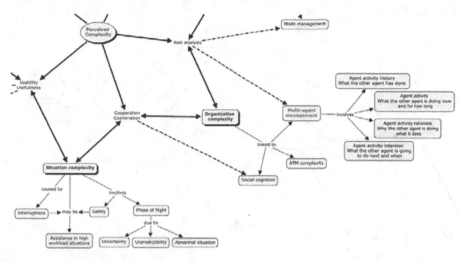

Fig. 7.9 Situation and organization complexity (CM-4)

saturation were proposed as typical indicators. Redundancy is always a good rule whether it repeats information for cross-checking, confirmation or comfort, or by explaining the "how", "where", and "when" an action can be performed. Ergonomics rules formalize user friendliness (i.e., consistency, customization, human reliability, affordances, feedback, visibility and appropriateness of the cognitive functions involved). Human reliability involves human error tolerance (therefore the need for recovery means) and human error resistance (therefore the existence of risk to resist to). To summarize, A-complexity analysis deals with the level of necessary interface simplicity, explanation, redundancy and situation awareness that a new artifact is required to offer to users.

Organization complexity (CM-4) is linked to social cognition, agent-network complexity, and more generally multi-agent management issues (Fig. 7.9).

There are four principles for multi-agent management:

- agent activity (i.e., what the other agent is doing now and for how long);
- agent activity history (i.e., what the other agent has done);
- agent activity rationale (i.e., why the other agent is doing what it does);
- agent activity intention (i.e., what the other agent is going to do next and when).

Multi-agent management needs to be understood through a role (and job) analysis. To summarize, an O-complexity analysis deals with the required level of coupling between various purposeful agents to handle the new artifact. Situation complexity is usually caused by interruptions and more generally disturbances. It involves safety and high workload situations. It is commonly analyzed by decomposing contexts into sub-contexts. Within each sub-context, the situation is characterized by uncertainty, unpredictability and various kinds of abnormalities. To summarize, a S-complexity analysis deals with predictability of various situations in which the new artifact will be used.

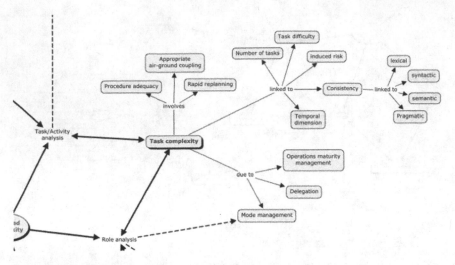

Fig. 7.10 Task complexity (CM-5)

Task complexity (CM-5) involves procedural adequacy, appropriate multi-agent cooperation (e.g., air-ground coupling in the aerospace domain) and rapid proto-typing (i.e., task complexity cannot be properly understood if the resulting activity of agents involved in it is not observable). Task complexity is linked to number of sub-tasks, task difficulty, induced risk, consistency (lexical, syntactic, semantic and pragmatic) and the temporal dimension (perception-action frequency and time pres-sure in particular). Task complexity is due to operations maturity, delegation and mode management. Mode management is related to role analysis. To summarize, a T-complexity analysis deals with task difficulty according to a spectrum from best practice to well-identified categories of tasks (Fig. 7.10).

Novelty Complexity, Creativity and Adaptation

Besides providing user requirements, users can be involved in the design process, especially in early stages, if mockups or prototypes are available. We must not forget that designers' and engineers' main asset is creativity. They are the ones who propose solutions. In addition, the human-centered design team needs to take the above dimensions into account to figure out the complexity of these solutions. Since maturity is at stake here, I claim that when high-level requirements are right from the beginning, subsequent developments, when they are carefully carried out, are not likely to lead to deep revisions when the artifact needs to be delivered. For that matter, user-perceived complexity needs to be tested from the very beginning of the design, when first ideas of the artifact are becoming drawable or writeable, and all along the design and development process.

In addition to being familiar with domain in which a new artifact will be tested, professionals who will have to analyze novelty complexity are required to have a clear awareness of and knowledge about the various relationships among user-perceived complexity and cognitive stability. Perceived complexity is more related to the "gulf of evaluation", and cognitive stability to the "gulf of execution", in Norman's terminology (Norman 1986). Even if adaptation is an asset of human beings, their lives are better when technology is adapted to them. Therefore, novelty complexity analysts need to better understand co-adaptation of people and technology in the perspective of increasing cognitive stability. Cognitive stability is defined by taking the physical metaphor of passive and active stability that respectively involves static and dynamic complexity. These concepts were taken into account to support human-centered design and have led to the following major principles: simplicity, observability and controllability, redundancy and cognitive support (Boy 2002).

Let us take a biological approach to understand complexity of interactive systems by finding out the salient parts and their interrelations. As already seen in Chap. 5, complexity is intimately related to separability. When a doctor administers a medication to a patient, he or she has to know the secondary effects of this medication (i.e., acting on a part may have an effect on other parts). When a part (e.g., the respiratory system), is failing, medication is usually provided to treat the lung disease, but this medication may have an impact on other parts of the body (i.e., the whole system). Of course, we will always attempt to separate what is separable in order to simplify! But there will be an end to this separability process. There are "atomic" parts that are not at all separable. These atomic parts live by themselves as a whole, eventually requiring interaction with other atomic parts. The problem is then to figure out how complex they are by themselves and what kind of complexity their interrelations generate. Designers and users of a system may not see the same parts and interrelations, just because they do not have the same tasks to perform, or the same goals to achieve, with respect to the system. They do not decompose the system in the same way because they do not have to understand the logic of the system in the same way. Separable parts and their interrelations can be seen as a conceptual model. The closest designer's conceptual model is to a user's conceptual model, the better. Therefore, people in charge of analyzing novelty complexity need to be aware of relevant parts and overall maturity evolution in terms of the AUTOS pyramid.

AUTOS-Complexity Criteria

Key criteria have been derived from the 62 elicited concepts[5] on novelty complexity presented on the above CMap. They were categorized to fit with the five AUTOS complexity sets of criteria that follow:

- A-complexity: interface simplicity, required explanation, redundancy and situation awareness.

---

[5] I realize the term "concept" can be confusing when we address both architecture and cognitive science communities. In architecture, a concept denotes the first drawings of a design. In cognitive science, a concept denotes something conceived in the mind, an abstraction, which is represented as a "term".

- U-complexity: user's knowledge, skills and expertise.
- T-complexity: task difficulty according to a spectrum from best practice to well-identified categories of tasks.
- O-complexity: required level of coupling between the various purposeful agents to handle the new artifact.
- S-complexity: predictability of the various purposeful situations.

These criteria may be dependent on each other. For example, analysis of the required explanation (A-complexity criterion) to handle a new artifact is a matter of maturity of the underlying technology. If it is mature, then complexity can be hidden from the user, otherwise it must be shown with the right level of required explanation. Consequently, a user needs to understand the provided explanation, and therefore have appropriate expertise (U-complexity criterion) and rely on current best practice (T-complexity criterion). Sometimes, a right coupling among actors dealing with the artifact (O-complexity criteria related to cooperation and coordination of activities for example) in predictable situations (S-complexity criterion) simplifies its usage. This example demonstrates the need for designers to master various categories of novelty complexity criteria and their possible interrelations.

Using novelty complexity criteria is a matter of either expert judgment or decomposition into indicators that enable designers to reach measurable variables. The former methods are usually called subjective, the latter are said to be objective. At some point, subjectivity always enters into the picture! Objective methods are either based on qualitative measures or require strong interpretation of quantitative results in the end. In order to facilitate the job of human-centered design teams, 63 indicators $\{I_j\}$ were developed from the elicited concepts, and related to novelty complexity criteria using a CMap. To summarize, a criteria C_i is a combination C_i of several indicators $\{I_j\}$. It is advised to run one or several brainstorming or GEM sessions, to determine appropriate combinations with domain experts. Such combinations can be modified along the analysis process as more knowledge is acquired on AUTOS-complexity of the new artifact being analyzed. The analysis is based on data acquisition methods varying from kinds of recording (e.g., parameters, verbal protocols and video) to observation, interviews and debriefings.

Modeling and Simulation Methods and Tools

Computer-aided design (CAD) is a software-based technique that enables designers and engineers to draft mockups and prototypes. It is initially geometry-based, but along the years it became full design support almost for any system, whether cars, aircraft, power plants, houses, kitchens, furniture or clothes. In the early days, people were drafting systems on paper using pens; today CAD supports the same kind of job using software. CAD now enables design and development of both structure and function of a system, bringing it to "life" by simulation. Even more important CAD led to development of the integrated design and development processes during the whole life cycle of a product (i.e., from first idea to design, manufacturing, delivery, marketing, operations, maintenance and dismantlement). CAD is also very important for documentation purposes, also during the whole life cycle of a product. Consequently in this section, we will talk about CAD not only as a drafting technique and tool, but also and mainly as a life cycle support system.

Drafting Objects

What is a design object? A design object is defined by three main attributes: a shape (or a structure), a behavior (or a function) and possible connections with other objects. There can be more attributes such as materials, processes, regulations related to its use and so on.

The first thing that we do to explain a complex idea is draw a picture! The more realist and accurate the picture is, the better we will be understood by our audience. We keep presenting statements, drawings and pictures on PowerPoint slides to make sure that our audience gets what we want to communicate. However, these presentations are not only for communication purposes, by making them we tend to rationalize what we think; the presentations we make are a kind of modeling and simulation of our purpose. CAD goes a step further by providing even more accurate features in space (2D and 3D), time (dynamics), and many other dimensions such as constituency, reliability and cost of various elements.

Today, CAD systems provide more than drawing capabilities. They provide possibilities for computer animation, online documentation, and traceability. For example, CATIA, developed by Dassault Systemes, is certainly a reference for CAD. Dassault developed CATIA internally in 1977 to support the development of the Mirage fighter airplane. Today, CATIA is used by many big manufacturers and suppliers allover the world.[6]

Integrating Objects into Systems

Design objects can be progressively integrated into engineered systems (i.e., designed tiles and rivets are progressively integrated into the wing of a spacecraft model). In addition, the local dynamics of a design object should be integrated into the design of a resulting embedding system. The "sum" of the dynamics of the various objects provides emerging dynamics to the embedding system. At this point, the **emerging behavior** can be either automatically generated by simulation, or directly programmed from an external source. Of course, we would like to be able to trust the modeling and simulation capability and discover emergent behaviors from the CAD system. This is not always the case, but this is a challenge that today's technology

---

[6] CATIA history by Francis Bernard (http://www.edstechnologies.com/download/history-catia. pdf).

might enable us to overcome in many cases. This is why understanding non-linear systems and complexity science is crucial in design.

Systems are often interfaced with other systems, but most importantly people have to interact with them. This is where **human-system integration** (HSI) enters into play, and the earliest the best! The first step of course, involves end users in the design of high level requirements using methods and techniques such as UML (Unified Modeling Language), but also human-in-the-loop simulations (HITLS) where end users will be able to generate emerging behaviors (we will present and discuss HITLS later in this chapter). The most important thing to remember at this point is that modeling and simulation techniques are important to rationalize HSI and enable the discovery of possible emerging behaviors to be further tested in real world experiments, like flight tests.

Integrating Systems into Systems

In the same way as objects are **integrated** into systems, systems are integrated into bigger systems, and so on. This is why the concept of systems of systems is so important to understand (Maier 1998; Carlock et al. 1999; Krygiel 1999; Sage and Cuppan 2001). In the military domain, the systems-of-systems approach has already been promoted mainly for enhancing interoperability and synergism.

Networks develop very fast on our planet to the point that the concept of systems of systems has become a tangible reality. The Global Earth Observation System of Systems (GEOSS) "enables us to envision a world where more people will be fed, more resources will be protected, more diseases will be mitigated or even prevented, and more lives will be saved from environmental disasters."[7] GEOSS is planned to be able to integrate data coming from many thousands of individual Earth observation technologies around the globe portraying various integrated ecological systems. GEOSS "architecture contains about 3,000 elements that are involved in earth science research: observation sources, sensors, environmental parameters, data products, mission products, observations, science models, predictions, and decision-support tools. The science models use observations from space-based instruments to generate predictions about various aspects of the environment. These predictions are used by decision-makers around the world to help minimize property damage and loss of human life due to adverse conditions such as severe weather storms. The architecture is developed using both traditional and nontraditional systems engineering tools and techniques" (Martin 2008).

In the first place, a system of systems (SoS) may not appear to be fully structured and functional. SoS development is evolutionary in the sense that functions and purposes are incrementally added, removed and modified with experience in the use of the system. Emergent behaviors will incrementally appear; some functions and purposes will be created, others will become obsolete, and others will merge or split.

---

[7] http://www.noaa.gov/eos.html.

This is an organizational learning process that will take place during the life cycle of the SoS; the SoS is intrinsically complex and adaptive. Each system in the SoS is independent and useful in its own right. Some systems may be redundant with others, and coordinated, to ensure the global stability of the SoS.

Integration of systems into a final system (e.g., aircraft systems into the aircraft itself) requires significant planning, preparation, time and resources. It is always the best solution to use a single facility for the integration, with competent people, organized and tested processes. Leadership is key; the on-site integration leader must be empowered for the operational community, supported by an SoS framework with sufficient resources and authority. A traceability process should be put in place and used effectively by competent personnel. Integration does not go without issues, last-minute problems and failures; contingency plans and schedules must be available. This is another reason to plan ahead during human-in-the-loop simulations of the integration process itself.

For all these issues and reasons, the Orchestra Model (developed in the Chap. 2) is entirely purposeful here. A common frame of reference (music theory) has to be set up for all actors to understand each other. Task assignments (scores) must be coordinated at the highest level (the composer). The integration leader must understand the overall SoS (by analogy with music, the symphony) and coordinate the actual integration with authority. Competent performers not only know their own disciplines perfectly, they also have a trans-disciplinary skill set and knowledge.

Discrete Event Simulation

A discrete-event simulation (DES) enables construction of a conceptual framework that describes the system (modeling), performing experiments using a computer implementation of the model (simulation), and draw conclusions from output that assist in subsequent decision-making (analysis). It is basically a chronological sequence of events that mark system states changes. The sequence of events is linear unless there is an abnormal situation that forces branching into another sequence of events. Generally, events are sequenced by a clock, which enables for a step from one event to another. DES can use several lists of events, which can eventually be chained appropriately (e.g., simulating nominal conditions or off-nominal conditions).

DES is attractive because it enables designers and modelers to compress or expand time (locally or globally), control sources of variation, avoid errors in measurement, stop and review, restore system state, replicate, and control the level of detail.

When you start a DES model, a few questions should come to mind, such as "is the system deterministic or stochastic?"; "static or dynamic?"; "continuous or discrete?" If the system is stochastic, some state variables are random. If the system is dynamic, time evolution is important. In the Monte Carlo simulation model for example, simulation is stochastic but time evolution is not important.

Another example can be the use of a discrete event simulation model to identify and understand the impact of different failures on the overall production capabilities in a chemical plant (Sharda and Bury 2008). What mattered in this case was to understand key equipment components that contribute towards maximum production loss and to analyze the impact of a change policy on production losses. DES was applied to enhance decision-making (i.e., a change policy in terms of new equipment installation or stock level increase for the failure prone components).

A DES model was also applied to study the impact of the effects of alternate team configurations and system designs on situational awareness in multi-unmanned vehicles control (Nehme et al. 2008). Such approach enables quantifying operator situational awareness by using data to more accurately predict metrics such as mission performance and operator utilization. In this case, DES allows us to avoid expensive, requiring time-consuming user studies.

DES is sometimes referred to time-line analysis (TLA), which are sequences of events are put on a time line. Time line analysis methods were developed and used to solve various problems such as aircraft cockpit certification (Boy and Tessier 1985). In human factors, TLA was initially conceived to better understand human operators' performance and workload. It is better used to identify changes that impact process performance. TLA is mostly used in accident analysis, but is also a very good modeling and simulation method for design.

Conclusion

Modeling and simulation is a crucial approach for the development and integration of human-centered design, which goes farther from its initial drafting intention. M&S can support the whole life cycle of a life-critical system from facilitating creativity and design thinking, to rationalization, information sharing, and finally evaluation and validation. It needs to be understood and managed at the highest hierarchical level of the organization. M&S is particularly interesting to analyze and design interaction among the various human and machine agents of an LCS, better understand its complexity, as well as its dynamics and control. M&S enables mastery of system novelty. M&S should not only be supported by computer-aided design, but also used for system integration; discrete event simulation and human-in-the-loop simulation as integrating parts of M&S.

References

Amalberti, R. (1996). *Controlling safety-critical systems* [in French]. Paris: Presses Universitaires de France.
Batten D. F. (2009). Changing our brains: Systemic causality in complex human systems. Proceedings of the 18th Conference on Behavior representation in modeling and simulation, Sundance, UT, 31 March–2 April, pp. 9–14.
Beer, R. D. (to appear). Dynamical systems and embedded cognition. In K. Frankish & W. Ramsey (Eds.), *The Cambridge handbook of artificial intelligence*. Cambridge: Cambridge University Press.

Boy, G. A., & C. Tessier (1985). Cockpit analysis and assessment by the MESSAGE methodology. In *Proceeding 2nd Conference on Analysis, Design and Evaluation of Man-Machine Systems* (pp. 73–79).

Boy, G. A. (1996a). *The group elicitation method: An introduction. Proceedings of EKAW'96*, Lecture Notes in Computer Science Series, Berlin: Springer.

Boy, G. A. (1996b). The group elicitation method for participatory design and usability testing. *Proceedings of CHI'96, the ACM Conference on Human factors in computing systems*, Held in Vancouver, Canada. Also in the *Interactions Magazine*, March 1997 issue. New York: ACM.

Boy, G. A. (1998a). Cognitive function analysis for human-centered automation of safety-critical systems. In *Proceeding CHI'98, the ACM Conference on Human Factors in Computing Systems* (pp. 265–272). Los Angeles: ACM Digital Library.

Boy, G. A. (1998b). *Cognitive function analysis*. Ablex Publishing, distributed by, Westport: Greenwood.

Boy, G. A. (2002). Procedural interfaces. In *Proceeding IHM'02 (the Francophone Conference on Human-Computer Interaction)*, Poitiers, France. New York: ACM Digital Library.

Boy, G. A. (2007). Perceived complexity and cognitive stability in human-centered design, *Proceedings of HCI International*, Beijing, China.

Boy, G. A. (2008). *Human-centered development of perceived complexity criteria: Developed criteria*. Technical Report DGAC/EURISCO No. T-2007-201. Direction Générale de l'Aviation Civile, Bureau des Aeronefs et de l'Exploitation, Paris, France.

Boy, G. A., Mazzone, R., & Conroy, M. (2010). The virtual camera concept: A third person view. *Third International Conference on Applied Human Factors and Engineering*, Miami, Florida; 17–20 July 2010.

Boy, G. A. (2011). Cognitive function analysis in the design of human and machine multi-agent systems. In G. A. Boy (Ed.), *Handbook of human-machine interaction: A human-centered design approach*. Aldershot: Ashgate.

Boy, G. A., & Grote, G. (2011). The authority issue in organizational automation. In G. A. Boy (Ed.), *The handbook of human-machine interaction*. Aldershot: Ashgate.

Brown, T. (2008). Design thinking. *Harvard Business Review*. http://hbr.org/2008/06/design-thinking/ar/1. Accessed 1 Sept 2012.

Brown, T. (2009). *Change by design: How design thinking transforms organizations and inspires innovation*. New York: Harper Business.

Caas, A. J., Ford, K. M., Novak, J. D., Hayes, P., Reichherzer, T. R., & Suri, N. (2001). Online concept maps: Enhancing collaborative learning by using technology with concept maps. *Science and Teachnology, 68*(2), 49–51.

Caas, A., Carff, R., Hill, G., Carvalho, M., Arguedas, M., Eskridge, T., Lott, J., & Carvajal, R. (2005). Concept maps: Integrating knowledge and information visualization'. In S. O.Tergan & T. Keller (Eds.), *Knowledge and information visualization: Searching for synergies* (Vol. 3426, pp 205–219). Heidelberg: Springer.

Cannon, W. B. (1932). *The wisdom of the body*. New York: W.W. Norton & Company.

Card, S. K., Moran, T. P., & Newell, A. (1983). *The psychology of human-computer interaction*. Hillsdale: Lawrence Erlbaum Associates.

Carlock, P. G., Decker, S. C., & Fenton, R. C. (1999). Agency level systems engineering for "systems of systems". *Systems and Information Technology Review Journal, 2*(Spring/Summer), 99–109.

Corker, K. M., & Smith, B. R. (1993). An architecture and model for cognitive engineering simulation analysis: Application to advanced aviation automation. In the *Proceedings of the AIAA Computing in Aerospace 9 Conference*. October, 1993: San Diego.

Coze, Y., Kawski, N., Kulka, T., Sire, P., Sottocasa, P., & Bloem, J. (2009). *Virtual concept: Real profit with digital manufacturing and simulation*. Dassault Systèmes and Sogeti. LINE UP Book & Media, The Netherlands, ISBN 978 90 75414 25 7.

Dreyfus, H. L. (1972). *What computers can't do: A critique of artificial reason*. New York: Harper and Row.

Dreyfus, H. (2002). Intelligence without representation: Merleau-Ponty's critique of mental representation: The relevance of phenomenology to scientific explanation. *Phenomenology and the Cognitive Sciences, 1,* 367–383

Endsley, M. R. (1988). Situation awareness global assessment technique (SAGAT). *Proceedings of the National Aerospace and Electronics Conference (NAECON)* (pp. 789–795). New York: IEEE.

Endsley, M. R. (1995a). Measurement of situation awareness in dynamic systems. *Human Factors, 37*(1), 65–84.

Endsley, M. R. (1995b). Toward a theory of situation awareness in dynamic systems. *Human Factors, 37*(1), 32–64.

Gibson, J. J. (1977). The theory of affordances. In R. Shaw & J. Bransford (Eds.), *Perceiving, acting, and knowing.* (ISBN 0-470-99014-7).

Gibson, J. J. (1979). *The ecological approach to visual perception.* Boston: Houghton Mifflin.

Grudin, J. (1994). Computer-supported cooperative work: History and focus. *Computer, 27*(5), 19–26.

Gunning, D., & Manning, M. (1980). The measurement of aircrew task loading during operational flights. *Proceedings of the 24th HFES Meeting* (pp. 249–252).

Harper, R. P. Jr., & Cooper, G. E. (1986). *Handling qualities and pilot evaluation. journal of guidance, control, and dynamics* (Vol. 9, pp. 515–529). Previously a conference given at the 1984 Wright Brothers Lectureship in Aeronautics.

Hart, S. G., & Staveland, L. E. (1988) Development of NASA-TLX (Task Load Index): Results of empirical and theoretical research. In P. A. Hancock & N. Meshkati (Eds.), *Human mental workload.* Amsterdam: North Holland.

Hart, S. G. (2006). *NASA-Task load index (NASA-TLX); 20 years later. Proceedings of the Human Factors and Ergonomics Society 50th Annual Meeting* (pp. 904–908). Santa Monica: HFES.

Heidegger, M. (1962). *Being and time.* Originally published in 1927. New York: Harper and Row.

Holland, J. H. (1998). *Emergence: From chaos to order.* Reading: Perseus Books.

Hutchins, E. (1995). *Cognition in the wild.* New York: MIT.

Javaux, D., & De Keyser, V. (1997). Complexity and its certification in aeronautics. *In the Proceedings of the 1997 IEEE International Conference on Systems, Man and Cybernetics,* Orlando.

Kauffman, S. (1993). *The origins of order—Self-organization and selection in evolution.* Oxford: Oxford University Press.

Kieras, D. E., & Polson, P. G. (1985). An approach to the formal analysis of user complexity. *International Journal of Man-Machine Studies, 22,* 365–394.

Kopardekar, P., Bilimoria, K., & Sridhar, B. (2007). Initial concepts for dynamic airspace configuration. *7th AIAA Aviation Technology, Integration and Operations Conference (ATIO).* AIAA 2007-776, Belfast, Northern Ireland, September.

Krob, D. (2011). Elements of complex systems architecture. Lecture given at the Florida Institute of Technology, February 7, 2011.

Krygiel, A. J. (1999). *Behind the wizard's curtain: An integration environment for a system of systems.* DoD C4ISR Cooperative Research Program, ISBN 1-57906-018-8.

Lakoff, G. (2004). *Don't think of an elephant!: Know your values and frame the debate—The essential guide for progressives.* White River Junction: Chelsea Green.

Luzeau, D., & Ruault, J. R. (Eds.). (2008). *Systems of systems.* Hoboken: Wiley. ISBN 978-1-84821-164-3.

Martin, J. N. (2008). Using architecture modeling to assess the societal benefits of the global earth observation system-of-systems. *IEEE Systems Journal, 2*(3), 304–311.

Maturana, H., & Varela, F. (1980). Autopoeisis and cognition: the realization of living. R. S. Cohen & M. W. Wartofsky (Eds.), *Boston studies in the philosophy of science* (Vol. 42). Dordecht: D. Reidel. ISBN 90-277-1016-3.

Maier, M. W. (1998). Architecting principles for systems-of-systems. *Systems Engineering, 1*(4), 267–284.

Merleau-Ponty, M. (1962). *Phenomenology of perception*. New York: Humanities.

Minsky, M. L. (1985). *The society of mind*. New York: Simon & Schuster.

Mitchell, M. (2008). *Complexity: A guided tour*. New York: Oxford University Press.

Myerson, R. B. (1991). *Game theory: analysis of conflict*. Cambridge: Harvard University Press. ISBN 978-0-674-34116-6.

Nehme, C., Crandall, J. W., & Cummings, M. L. (2008). Using discrete-event simulation to model situational awareness of unmanned-vehicle operators. *Proceedings of the ODU/VMASC Capstone Conference*.

Norman, D.A. (1986). Cognitive Engineering. In D. Norman & S. Draper (Eds.), *User-centered system design* (pp. 31–61). Hillsdale: Lawrence Erlbaum Associate.

Norman, D. A. (1988). *The design of everyday things*. New York: Doubleday.

Prevot, T., Mercer, J. S., Martin, L. H., Homola, J. R., Cabrall, C. D., & Brasil, C. L. (2010). Evaluation of NextGen air traffic control operations with automated separation assurance. Presentation at the International Conference on Human-Computer interaction in Aerospace Conference (HCI-Aero), West Lafayette, In: Human-Computer Interaction International.

Prigogine, I. (1997). *The end of certainty*. New York: Free.

Pylyshyn, Z. W. (1985). *Computation and cognition. Toward a foundation for Cognitive Science*. Cambridge: MIT.

Rasmussen, J. (1986). *Information processing and human-machine interaction—An approach to cognitive engineering*. Amsterdam: North Holland.

Rauzy, A. (2008). Guarded transition systems: A new states/events formalism for reliability studies. *Journal of Risk and Reliability, 222*(4), 495–505. Professional Engineering Publishing.

Reason, J. (1987). Cognitive aids in process environments: prostheses or tools? *International Journal of Man-Machine Studies (Special Issue): Cognitive Engineering in Dynamic Worlds, 27*(5–6), 527–539.

Robinson, S. (2004). *Simulation: The practice of model development and use*. New York: Wiley.

Ruelle, D. (1989). *Elements of differentiable dynamics and bifurcation theory*. London: Academic. ISBN 0-12-601710-7.

Sage, A. P., & Cuppan, C. D. (2001). On the systems engineering and management of systems of systems and federations of systems. *Information, Knowledge, and Systems Management, 2*(4), 325–345.

Sharda, B., & Bury, S. J. (2008). A discrete event simulation model for reliability modeling of a chemical plant. In S. J. Mason, R. R. Hill, L. Mönch, O. Rose, T. Jefferson & J. W. Fowler (Eds.), *Proceedings of the 2008 Winter Simulation Conference* (pp. 1736–1740). Piscataway: Institute of Electrical and Electronics Engineers, Inc.

Sollenberger, R. L., Willems, B., Della Rocco, P. S., Koros, A., & Truitt, T. (2005). Human-in-the-loop simulation evaluating the collocation of the user request evaluation tool, traffic management advisor, and controller-pilot data link communications: Experiment I—Tool combination. Technical Note, DOT/FAA/CT-TN04/2.

Stephane, A. L. (2011). Eye tracking from a human factors perspective. In G. A. Boy (Ed.), *The handbook of human-machine interaction: a human-centered design approach*. Aldershot: Ashgate.

Straussberger, S., Chamayou, C., Pellerin, P., Serres, A., Salis, F., Feuerberg, B., Lantes, J. Y., Guiost, B., Figarol, S., Reuzeau, F., & Boy, G. A. 2008. *Scenarios in PAUSA*. DGAC-DPAC EURISCO, Technical Report. April.

Suchman, L. A. (1987). *Plans and situated actions*. Cambridge: Cambridge University Press.

Taleb, N. N. (2007). *The black swan: The impact of the highly improbable*. New York: Penguin. ISBN 978-1-4000-6351-2.

Thesen, A., & Travis, L. E. (1989). Simulation for decision making: An introduction. In E. A. MacNair, K. J. Musselman & P. Heidelberger (Eds.), *Proceedings of the 1989 Winter Simulation Conference* (WSC'89). New York: ACM Digital Library. ISBN:0-911801-58-8.

Van Daele, A. (1993). *Complexity reduction by operators in continuous process control* [in French]. In Work Psychology Department, Liège: Université de Liège.

Varela, F. J., Thompson, E., & Rosch, E. (1991). *The embodied mind: Cognitive science and human experience*. Cambridge: MIT.

Vicente, K. J. (1999). *Cognitive work analysis: Towards safe, productive, and healthy computer-based work*. Mahwah: Lawrence Erlbaum Associates.

Winograd, T., & Flores, F. (1986). *Understanding computers and cognition: A new foundation for design*. Norwood: Ablex.

Xing, J., & Manning, C. A. (2005). Complexity and automation displays of air traffic control: literature review and analysis. Civil Aerospace Medical Institute, US DOT Federal Aviation Administration, Oklahoma City, OK 73125, April 2005, Final Report. This document is available to the public through: (1) The Defense Technical Information Center, Ft. Belvior, VA 22060; (2) The National Technical Information Service, Springfield, VA 22161.

Chapter 8
Advanced Interaction Media

Introduction

User interfaces are recognized as the major part of work for real programs, approximately 50 %. They have a very important impact on the way products are used, make the reputation of the organization that produces them, and may cause safety issues. Usability engineering (Nielsen 1993) has become a mandatory discipline in the production of systems. According to the International Standard Organization (ISO), usability is "the effectiveness, efficiency, and satisfaction with which specified users achieve specified goals in particular environments" (ISO 9241-11).[1] This being said, we need to be careful about the interpretation of this definition. Indeed, "users" can be novice, occasional or expert in a given domain.

However, the concept of user interface will become progressively of lesser importance because user interface will be designed as an integrating part of the overall system. Instead of the conventional inside-out approach of engineering where the user interface is designed in the end, the outside-in approach of human-centered design promotes participation of actors dealing with the system being designed. Design is deeply grounded in communication, expertise and experience sharing, creativity and incremental rationalization.

Since software is now everywhere, in various kinds of appliances, buildings, transportation means, industrial processes and at the heart of our homes, we need to learn how to interact with these new objects. In addition, computer networks support interaction among people, and our fundamental ways of interacting between each other must be re-learned. New kinds of interaction media are all around us, and this evolution process is far from being stabilized. This is why I think it is important to focus on what I call "advanced interaction media".

Human-Computer Interaction has become a field of research and part of industry. It was born in the early eighties. Engelbart is certainly the father of technology that supports HCI and collaborative technology[2] today. He invented the mouse and

---

[1] http://www.iso.org/iso/catalogue_detail.htm?csnumber=16883.

[2] The term "computer-supported cooperative work" (CSCW) was coined in 1984 by Paul Cashman and Irene Grief to describe a mutlidisciplinary approach focused on how people work and how

was one of the engineers who worked on the ARPANET project in the 1960s. He was among the first researchers who developed hypertext technology and computer networks to augment the intellectual capacities of people.

During the late seventies and even more during the eighties, office automation was born from the emergence of new practices using minicomputers. Minicomputers and microcomputers were integrated in many places such as travel agencies, administrations, banks and homes, to support individual, groups and organizations. People started to use them interactively, as opposed to using them in batch mode. Single user applications such as text processors and spreadsheets were developed to support basic office tasks. Several researchers started to investigate the way people were using this new technology.

Computer science was originally the science of internal functions of computers (i.e., how computers work). With the massive use of computers and their incremental integration in our lives, computer science has also become the science of external functions of computers (i.e., how to use computers and what they are for). Cognitive and computer scientists needed to investigate and better understand how people individually and collectively appropriate computers to support collaborative work. Multi-disciplinary research was developed by involving psychologists, sociologists, education and organization specialists, managers and engineers.

People work together using various kinds of support to communicate, cooperate and coordinate their activities. Paper documentation, telephone, electronic mail, Internet, Intranets, mobile computing, and desktop conferencing are examples of such supports. People communicate, share information and coordinate activities synchronously or asynchronously, in the same place or remotely.

The Prolific Sixties

The sixties were rich in many ways. The Apollo program, ARPANET and the early start of hypertext were concomitant and totally independent at the same time. Apollo brought the first man to the Moon. ARPANET brought computer network technology. Hypertext brought revolutionary nonlinear capabilities to navigate through information. What do these innovations and discoveries have in common?

If we go back to the Moon in the future, technology will be different than in the sixties. However, this program concretely introduced digital computers into spacecraft and later on into aircraft. David A. Mindell wrote a magnificent book, *Digital Apollo*, and I would like to share his description of how digital computers were introduced in spaceflight. "Tying the whole thing together was an embedded digital computer, made out of exotic devices called "integrated circuits"—silicon chips, running a set of esoteric programs. In the middle of the instrument panel, amid familiar dials and

---

technology could support them. CSCW scientific conferences were organized in the USA within the ACM-SIGCHI community (Association for Computing Machinery—Special Interest Group on Computer Human Interaction). Conferences on the topic immediately followed in Europe and Asia. Related work and serious interest already existed in European Nordic countries.

switches, stood the computer interface, a numeric keypad glowing with segmented digits. Throughout these space missions the astronauts punched in numbers, ran programs, and read the displays. Much of the landing was under direct control of these programs. When he flew, Neil Armstrong did not command the spacecraft directly, but rather used two control sticks to command the computer, whose programs fired the thrusters to move the Lunar Module. Every move was checked and mediated by software, written by a group of young programmers half a world away." (Mindell 2008, pp. 1–2). The future of flying was right there. We had to wait until the eighties to see the first "glass cockpits " and "fly by wire" technology on commercial aircraft.

Computer technology does not solve problems. During the Apollo 11 mission, suddenly the Lunar Module lost contact with the Earth mission control center because the main antenna had problems. The crew had to turn off the automatic control in order to fix the antenna by hand. Fortunately, they solved the problem correctly just like in simulations done prior to the mission. Scenario-based training was and is key. Many of these events happen and need to be anticipated as much as possible; this would be impossible without a human-centered approach. Before landing on the Moon, Neil Armstrong saw a large crater where they were supposed to land. He then decided to land manually. This was based on his strong knowledge of the Lunar Module systems and their limitations. Competence of human operators is key in life-critical systems. Therefore, computers in the cockpit are fine and very helpful, but they do not eliminate pilot skills and domain knowledge. Furthermore, they require skills and knowledge on software capabilities and limitations.

Beside training and competence, the Apollo 11 crew benefited from the communications with the mission control center as well as various kinds of skills and knowledge from specialists on the ground. Interconnectivity was experienced like it has never been in the past, if we take into account the distance. Today distance between people is no longer an issue with the growing sophistication, availability and user-friendliness of computer networks. Even if astronauts of the Apollo missions did not benefit from the nascent Advanced Research Projects Agency Network (ARPANET) developed by the Defense Advanced Research Projects Agency (DARPA) of the United States of America (US) Department of Defense (DoD), it is interesting to notice the concomitance of both programs. ARPANET's goal was to connect DoD's projects at universities and research laboratories in the US (Marill and Roberts 1966). ARPANET was the pre-figuration of Internet.

In writing this book, I have been using the Internet constantly to verify concepts that I am using. The Internet today provides a tremendous amount of flexibility and power that is at our fingertips. Not only searching for information on various kinds of databases, I also used Internet-based voice communication to ask some of my colleagues for advice on difficult topics. This would not have been possible without mastering ARPANET protocols such as e-mail and FTP (file transfer protocol) in the seventies. We needed to wait a little bit longer for voice over Internet protocol (Leiner et al. 1997).

Why does the Internet works so well? Maturity of the technology is an answer, but this is not sufficient. Maturity of practice is another even more important answer. People do not have any problem using the Internet because it is user-friendly. As we use this tool everyday, we tend to forget this was made possible thanks to the integration of hypertext into network protocols (i.e., going from FTP to HTTP, the hypertext transfer protocol). Tim Berners Lee was recognized for this accomplishment when he was working at CERN[3] in Geneva, Switzerland. This was in the beginning of the nineties. However we need to remember that hypertext started in the sixties, with Douglas Engelbart's and Ted Nelson's work in particular. Remember that Engelbart was also involved in ARPANET. All these people understood that data is comprised of relationships. That means linking people and data, but also people and people through smart information links. Today, data appears in various forms, including vivid ones form provided by real people wherever they are in the world.

Ubiquitous Computing

This fantastic evolution of technology went through the apparition of personal computers and the desktop. Office automation, like cockpit automation, happened at the same time, during the nineteen eighties. Of course we had autopilots in commercial airplanes since the thirties, but the real major step from an information technology perspective occurred with fly-by-wire technology. Office automation provided necessary tools to master all kinds of tasks that were typically done in offices. Text processing was a big step. I often use the *Knowledge Navigator*,[4] a short clip produced by Apple in 1988, depicting future activities in an office, to start my graduate student discussions on how accurate these projections were, except for natural language understanding. Today, we find it "natural" to type text and illustrate with pictures we find on the Internet, talk with other people at the same time on the same machine, and so on. Software becomes smarter and smarter, enabling automatic typo corrections and learning from our past searches for example.

How do we make ubiquity? First, we need to make computing mobile and connected. Second, we need to instrument the person. Third, we need to instrument the physical environment. Ubiquitous computing is pervasive (i.e., embedded and transpiring in "everything" we do); it can be wearable. When correctly done, this type of computing may lead to intelligent environments. Augmented reality is a good example of ubiquitous computing. A pervasive environment is saturated with computing and communication capability. Computing should be gracefully integrated with and disappear from its users. Current smart phones belong to this category of device, especially those equipped with multi-modal inputs and outputs. They also should be robust and reliable (i.e., pre-configured and with very rare failures). Of course, this is a matter of technology maturity, as well as maturity of practice.

From a human-centered point of view however, the most impressive step was ubiquitous computing, coined by Mark Weiser in 1988, when he was Chief Technologist at Xerox PARC. He recognized that designing machines that fit the human

---

[3] Centre Européen pour la Recherche Nucléaire (European Organization for Nuclear Research).

[4] http://www.youtube.com/watch?v=QRH8eimU_20.

environment requires understanding psychological, social and cultural phenomena. Weiser made the point that computing systems will become less intrusive and fade into the background of our everyday lives. From a philosophical standpoint, ubiquitous computing requires a shift from positivism to phenomenology. In other words, even if we need to cut things into pieces to engineer machines, we also need to keep a holistic view of the end product, and that involves a multitude of possible interactions.

According to Weiser, we are freed to use artifacts without thinking only when they disappear. It is the case for example, with car engines, that current drivers forget and only concentrate on driving instead of worrying about mechanical aspects of their engines, as in the early days of the car industry. Today, we enter into a room and it lights up without any specific human action except the fact that we move into a beam that triggers an appropriate mechanism. Not only is this function comfortable, it is also energy efficient. When we go out, a system turns the light off automatically. We forget the lighting mechanism and we can concentrate on what we have to do. Of course, there are invisible computers in the background (Norman 1998). "Rather than turning inward to an artificial world, ubiquitous media encourages us to look outward. It expands our perception and interaction in the physical world." (Buxton 1997).

Ubiquitous computing, or Ubicomp, typically refers to natural interfaces, context-aware applications, and automated capture and access (Abowd and Mynatt 2000). Ubicomp enables moving from discrete interaction toward continuous interaction, transforming engineered human-computer interaction into human and computer symbiosis. A user interface becomes natural when the user forgets it exists. Context-aware systems are necessarily adaptive taking into account sensory information and transforming it into a response (e.g., the light that turns on when you enter the room). Automated capture of information is not a common practice, usually called a trace that can be accessed later on. Also, huge quantities of information such as in YouTube or Facebook are being captured everyday, and are available to anyone. This is both very convenient and terrifying, because such information cannot be forgotten. Ethics is at stake here.

The main problem in context-aware technology is the definition of context. Context is very difficult to grasp since it can entail time, space, various kinds of environmental conditions, modalities of use and so on. Being aware of a situation is not only a matter of data gathering; it is also a matter of interconnections between these data. In other words, situation awareness is linked to a dynamic complex model, which enables perception, interpretation, understanding and follow-up action. Understanding the current context is a matter of expertise and experience. People who are expert and have extended experience are better than anybody else in assessing a situation and understanding what to do. Consequently, context-aware technology will always be specific to very well-known situations.

Ubiquitous computing is great, but it also can alienate us. Indeed, we are and need to be constantly wired to support the system that is intended to support us. Computers are being embedded everywhere in the environment and in appliances of every kind. Computers are disappearing but they are omnipresent allover our environment and constantly available. In order to be effective, computers will be sensing more of everything we do or say, rather than what we type. This of course introduces issues of privacy, trust and security (Langeheinrich 2001).

Dourish (2001) studied the concept of "embodied interaction" that reflects phenomenological approaches of Martin Heidegger, Ludwig Wittgenstein and other twentieth-century philosophers. He focuses on human-computer interaction that emphasizes skills and engagement, putting to the front the phenomenological tradition that claims primacy of natural practice over abstract cognition in everyday life. Dourish shows how this perspective can shed light on the foundational underpinning of current research on embodied interaction. He looks in particular at how tangible and social approaches to interaction are related, how they can be used to analyze and understand embodied interaction, and how they could affect the design of future interactive systems.

From Paper to Electronic ...

We have focused on electronic procedures on board spacecraft for quite a while. During the golden eighties when expert systems were blooming, circa 1984 we developed the Human-ORS-Expert-System (HORSES) for the support of Orbital Refueling Systems (ORS) operations on board the Space Shuttle (Boy 1986, 1987). This electronic operational documentation was supporting refueling operations in life-critical situations. While developing and testing these operations, we realized that computer support could be superior to paper in many situations. Most importantly, computer support could be connected to the systems being monitored and controlled. This provides contextual indexing capabilities, and consequently faster information retrieval. We took this research result and hence we designed the Computer Integrated Documentation (CID) that led to contextual indexing (Boy 1991a,b). Indeed, we proved that we could develop a machine learning mechanism that improves contextual indexing, typically implemented on hyperlinks between descriptors (i.e., mostly in the form of keywords) and referents (i.e., target documents). At the same time, the Web was proposed by Berners-Lee (1990) who invented it and developed its first versions. Development of the semantic Web followed (Berners-Lee 2001).

The first e-books were designed and developed during the mid-sixties, with the NLS project at Stanford Research Institute (Engelbart 1966) and Hypertext Editing System project at Brown University (Van Dam 1970). It took more than 3 decades to see consistent and mature use (e.g., Kindle developed by Amazon and Nook developed by Barnes and Noble). Again, the main problem was to find out what the main emerging properties of the use of e-books are. Why is this technology better than paper books? Why and when paper is still a great resource? In what contexts is electronic better than paper? We are still trying to answer these questions, not because we did not find rationale for both solutions, but because we can do more things than before. For example, tangible interaction with software-intensive appliances enables combining electronic and paper capabilities. For example we now have paper-based passports that are used as before, but they also have RFID[5] capabilities that enable airport police to see more than what is actually visible on the paper passport itself. This kind of capability brings to the front a distinction between public and private information, and who can see what.

Recently, newspapers have converted their paper-based information into computer-based information. It is now common to read news on the Web either on your computer, smart phone or tablet. Information is not formatted in the same way both to ease reading and to enable hypertext navigation. Since most people are equipped with appropriate technology, paper-based newspapers tend to disappear or at least decrease in volume. This is good for our trees! However, we are in the middle of the river; there are several reasons why paper-based newspapers and journals are superiors to electronic information, and actually used. First, people find that newsprint is tangible, we can hold it and use it using real pens for annotations and other things. Again, this could be interpreted as a technology maturity issue. Second, reading newspapers on the web gets distraction from other applications all the time; this is a maturity of practice issue. Third, people do not need to be connected to a WIFI network, which is commonly the case in many locations; this is another technology maturity issue. Fourth, it is enjoyable to buy a beautiful magazine and actually keep it on a shelf for further reading and sharing with family or friends in a quiet atmosphere.

The Web has become an external memory for everybody, to the point that we now get lost when we do not have access to it. Electronic connection is a reality in our everyday life. For example, it is used for exploration, retrieval and verification. It is a cognitive support, sometimes a cognitive prosthesis. I think the Web makes us smarter as long as we keep our educated common sense and critical capabilities (i.e., we are able to "buy" things that make sense). The Web makes us smart because we do not have to remember details, and allows us concentrate on ideas, creativity and rationalization. This again requires maturity of practice.

Tangible Interaction

More generally, not only are there things that require continuity with the past, we also need to investigate what kind of technology requires tangible interaction. There are three main cases: dual use; meaningful manipulation; and embodiment. Dual use is when tangible interaction involves physical interaction (e.g., somebody reading information on a paper passport), and virtual interaction (e.g., a machine reading RFID information on a passport). The passport remains a tangible support of public information, but it also has virtual embedded capabilities. In both cases, technology vanishes in the background to provide natural interaction. Meaningful manipulation is often required when situation awareness is at stake. In other words, interaction

---

[5] Radio Frequency Identification.

should make sense sometimes not only at the conscious level, but also at the sub-conscious level where an affordance relationship is built between technology and people. Finally, interaction with technology should be embodied (i.e., technology emphasizes skilled, engaged practice rather than disembodied rationality). Embodied phenomena occur in real-time and real-space. Thus we are back to ubiquitous computing.

If technology wants to be forgotten and leave primacy to natural interaction, it has to be robust, reliable and resilient in case of failure ; otherwise human operators are required to know about it (i.e., how it works and it should be manipulated). This is still a big question in aviation: what should pilots know about flying and avionics systems? The answer to this question is grounded in the flying activity. Since flying is not natural to people, it mainly remains a highly cognitive activity. For that matter, pilots should know about both flying (i.e., airmanship) and avionics systems, plus know how to follow procedures. In particular, the conventional desktop interface may not be the right metaphor to deal with the very procedural work that a commercial pilot has to do. We proposed the "procedural interface" solution, which is more appropriate for this kind of work in routine situations (Boy 2002). I would easily generalize this statement to life-critical systems for both procedural support to routine work and expert competence for handling the unexpected.

"We encounter the world as a place in which we act" (Heidegger 1962). We have seen that context is an important factor in human-computer interaction, and more generally in the socio-cognitive stability of the TOP model (i.e., technology, organization and people). Tangible interaction takes place in a tangible social context. Very recently, social computing emerged as a reality through groupware applications, including the Web. Consequently, new kinds of factors such as **accountability** and abstraction also emerged from the use of such technology. When people are interacting in a social context, they should be accountable (i.e., their actions should be observable and reportable). Interactions should be evident to people in context. In fact, action and accountability cannot be separated. People should share a common sense of understanding. We are back to the Orchestra Model where a common framework (i.e., a music theory) should be shared to enable mutual understanding, or by default a mediating representation that guarantees safe, efficient and comfortable interaction. Emphasizing **intersubjectivity**, Schultz (1973) assumed that given our experiences of the world are our own (i.e., they are subjective), we try to achieve a common experience of the world with other people, and then build a shared framework for meaning. Today, we are reconstructing new intersubjectivity via new technology and organizations.

We need to also acknowledge that software systems are built from **abstraction**, which tends to hide information, as well as how actions are organized. The question is: what should the designer show in the interface? What should the level of granularity be, that is shared with users or human operators? All actions are not relevant to share. However, some are and they need to be identified. For example, the already described "interactive cockpit " has become less tangible than its predecessors, which were interactively more tangible. For the last 30 years, we have embedded software into cockpits and aircraft to improve their manipulation, but pilots are still in charge of flying aircraft. Therefore, cockpits should be tangible. The question is: what level of delegation should we put in tangible interfaces? How much of this delegation can be understood by human operators? What is the level and nature of proficiency that human operators should have? Even indirectly, pilots have to interact with physical artifacts such as wings, fuel, flaps, slats and landing gear. They also have to interact with abstract artifacts such as thrust and lift. Finally, they have to interact with natural things such as wind, weather and relief. Information technology enables the development of intertwined meaningful, embodied and ubiquitous indices of the reality between people and machines.

During the European DIVA[6] project (1998–2001), we studied a navigation display (ND) that integrated a vertical profile under the existing horizontal display (Boy and Ferro 2000). It was interesting to notice that pilots preferred 3D navigation displays to 2D navigation displays because they found it more natural. However, we also noticed that they were so focused on such 3D display that they sometimes forgot to focus on the Primary Flight Display (PFD), which is (by definition) primary to the flight. The main problem was that it was impossible to integrate these two displays (i.e., ND and PFD). Therefore secondary instruments with highly attractive features and affordances may distract the overall performance of human operators.

Advanced Interaction Media for Product Management

A few important properties of advanced interaction media (AIM) have been described above in this chapter. These descriptions were mainly provided for products embedded with software technology. This section is devoted to the way products can be managed during their life cycle, and AIM is a great support for such product management.

First, life-critical systems are not designed and developed by a single individual. They are built by teams and often by large organizations. Their life-criticality requires support such as explicit rationale for the main processes such as design, traceability, certification, training, maintenance and decommissioning. Consequently, useful and usable documentation has to be produced and used. At the same time, we should be aware that current practice of quality assurance creates problems because employees spend too much time writing reports to show good practice ("I did my job well!") to the point that reporting is often more important that doing the work itself. Consequently, it is important to provide appropriate technology and organization setup that facilitate interconnection between the various actors of a team, organization or community. It is also important to train employees to effectively communicate among each other.

---

[6] Design of Interaction and Validation in Aeronautics.

Active Design Documents and Storytelling

What kind of information should we keep about the genesis and evolution of a product? There are four main kinds of information that are important to keep during the life cycle of a product. First, design rationale evolution provides decision-making processes that led to the selected solutions. Second, solutions themselves should be explained in terms of structures and functions (i.e., how do these solutions work?). Third, the way solutions should be and/or are effectively used (i.e., what are the tasks and activities related to the solutions?). Fourth, various evaluations should be documented in terms of usefulness and usability factors. The active design document (ADD) representation captures this kind of description (Boy 2005). Of course, there are many other ways of representing useful information for management of the life cycle of a product (Stark 2004, 2007; Saaksvuori 2008).

Why should we keep a contextualized trace? During the mid nineties, my research institute, EURISCO, carried out a knowledge management study on Concorde's life cycle because even if technical information on this aircraft was available in the form of reports, engineers had difficulty understanding it. **Vivid knowledge** was necessary to improve their understanding. In order to be used effectively, vivid knowledge requires extensive description of the context in which this knowledge was produced. Very often, context is made of episodes that only the people who generated knowledge know. In other words, vivid knowledge is always associated with a story. Unfortunately in the case of the Concorde's study, people who generated the reports and could tell associated stories were either retired or no longer alive. Consequently, it is crucial to contextualize and preserve technical information produced during the life cycle of a product.

Context is difficult to capture and may have various kinds of structures. As already said, context is very well captured from storytelling. Telling the right story well is an art that requires involvement. Since there is no human involvement without motivation, it is crucial to maintain high motivation in organizations. This is where creativity comes back, as well as the need for aesthetics in engineering activities. Even if the ADD representation is very important to rationalize captured knowledge, knowledge sources and content, the way this knowledge will be expressed and linked are even more important. Appropriate links between knowledge pieces constitute an effective way to handle context.

Facilitating Interaction in a Broader Sense

Design is not just making things. Telling the right story that illustrates a piece of knowledge is the right way to go in HCD. How? Being a good storyteller requires skills that are both an art and grounded on specific techniques. Storytelling is a powerful way to bring a concept to life. Indeed, design is a social activity that should bring together team members in a collaborative way. Communication is key. This is why we need tools that enable us to correctly support and even mediate interactions among team members, whether they are humans or machines (i.e., software-based tools or robots).

Current technology enables us to express things in a more direct and natural way. Indeed, it is typically better to show an animation of a specific design using an appropriate scenario than to explain this design with (a thousand) words. Currently M&S is very powerful for handling static objects. Advanced interaction media are providing the necessary dynamics to transform current M&S into appropriate animations.

M&S needs then to be understood as a tool and a technique supporting participatory design of dynamical complex systems. The "interaction" dimension has become essential. First and foremost, interaction requires interaction devices. Pointing devices (e.g., mouse) are now commonly used by a wide range of people. We know the pointing tasks such as "Select" (e.g., menu selection), "Position" (i.e., choose a point), "Orient" (i.e., choose a direction), "Path" (e.g., curving a line in a drawing program), "Quantify" (e.g., specify a numeric value), "Gesture" (e.g., a rapid back and forth motion to erase), and "Text" (i.e., enter, move, edit text in a 2D space). Pointing device control could be done directly, such as with the lightpen, touchscreen and stylus. It also can be done indirectly, such as with a mouse, trackball, joystick, trackpoint, touchpad, and graphic tablet. There are special purpose pointing devices such as multi-touch tablets and display, bimanual input, eye trackers, sensors, 3D trackers, data Gloves, boom Chameleon, haptic feedback, foot controls, tangible user interfaces, and digital paper. Of course, each of these devices is more or less appropriate for a specific use, and is evaluated with respect to principles and criteria such as speed and accuracy, efficacy of task, learning time, cost and reliability, size and weight. Fitts's law can be useful to determine the time it may take to reach a target using a pointing device (Fitts 1954):

$$MT = a + b \log_2 \left(\frac{D}{W} + 1 \right)$$

where MT is the movement time, a and b are empirically derived constants, D is a distance to the target, W the size of the target. MT increases as the distance to the target increases, and decreases as the size of the target increases. For example, eye tracking is superior to hand, mouse, joystick, trackball and touchpad (Ware and Mikaelian 1987; Fitts 1954; Card et al. 1978; Epps 1986; MacKenzie 1992).

Keyboard and cursor control units (KCCU) were recently introduced in new commercial aircraft. This new feature contributed to development of the concept of **interactive cockpit**, which itself induces several distinction issues such as continuous versus discrete interaction, and concrete versus virtual interaction. Indeed, in the past cockpits were interactive, but for very different reasons. Traditional cockpits are interactive because pilots are able to interact with the mechanical parts of the aircraft, using control columns that were replaced by joysticks in some commercial airplanes during the eighties. Joysticks, or side sticks, introduced a new way of interacting with the mechanical parts of the aircraft through electronic systems. More generally, computer systems have been introduced in aircraft to the point that flying is replaced by system management most of the time. In order to manage these systems, software-based interaction devices needed to be introduced, and the modern cockpit has become an interaction space in the human-computer interaction sense. Consequently, pilots now interact more with computers (i.e., involving discrete and virtual interactions) than directly with mechanical parts of the aircraft (i.e., involving continuous and concrete interactions).

Interactive cockpits keep some kinds of ergonomic continuity with their ancestors for basic functions, such as flying qualities. However, the new possibilities provided by software increases number of functions related to systems. Pilots are accountable for the overall performance and safety of the aircraft, but they have additional tasks related to management of systems performance. This created a new job that pilots had to integrate in their everyday activity. A new evolution of their job will happen with integration of air traffic management in cockpits.

AIM Facilitates Product Integration

We already saw that HCD put forward an outside-in approach of the architect . How can this be implemented? Let us draw a picture of what is going on today. Everything is decided from the top down moneywise, but technology is generated bottom up by suppliers in a very anarchical way. Governance is financial and not socio-technical. How on this condition can the outside-in approach be implemented and successful? It just cannot. Architects need reclaim governance. We need architects who are composers and able to design successful symphonies. These architects need appropriate support that today, relies on advanced interaction media associated with modeling and simulation. This is support for product integration during the whole life cycle of a product.

Choosing the right AIM is crucial to facilitate communication among the actors in the life cycle of a product. It starts by interactive M&S used to design the first mockups and prototypes: these prototypes are used to test usefulness, usability, reliability, resilience and finally maturity of both technology and associated practices. Examples of such AIM are post-its, paper-and-pencil prototypes, rapid prototyping, and more advanced computing environments associating M&S, interaction and domain dynamics. These AIM systems must involve human-in-the-loop possibilities. The more you test, the better a product will be adapted to high-level requirements.

Computer-supported cooperative work (CSCW), as an AIM, promotes communication, cooperation and coordination of agents who use them (Grudin 1994). Examples of such AIM systems are email, web-based communication systems (e.g., Skype and Webex), and shared organizational memory systems (e.g., DOORS). They now provide a framework for collaboration activities to support teams, small groups, organizations, as well as communities and e-commerce.

AIM Is Part of the Product

Commercial aircraft designed today will include more AIM (e.g., bigger interactive screens, information navigation tools and integrated communication systems with

Fig. 8.1 The classical cycle of documentation in industry

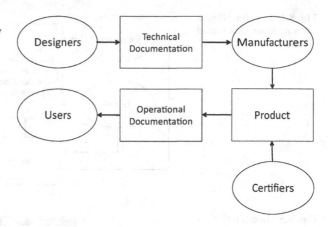

the ground), and include web capabilities. We are prototyping navigation systems (e.g., the virtual camera) for spacecraft that include multi-touch displays, voice input and tablets with accelerometers, for example.

Therefore, what is the best way to integrate AIM in a product? AIM should facilitate interaction between users and the product. There are three complementary ways to do this: design a usable user interface; design appropriate operational procedures; and correctly train people who will use the product. I have already shown and discussed the **duality of user interface and operational procedures** (Boy 1998). In reality, procedures are often designed and developed when systems are fully developed, and it is always too late (i.e., they are then used to compensate some user interface design flaws instead of supporting human-machine interaction). More generally, the more transparent a user-interface is, the less it demands procedures to use it. Conversely, the more opaque a user-interface is, the more it demands procedures and guidance to use it. In addition, people seldom use procedures when they are not obliged to do it (i.e., it is not a primary reflex).

The classical cycle of documentation in industry is represented in Fig. 8.1. Designers and engineers develop technical documentation that describes the various structures and functions of the product, and later on, when the product is developed and certified, the operational documentation is developed. This is what I called the inside-out approach to systems engineering .

More organizations have improved this classical cycle of documentation by using technical documentation as a shared database (Fig. 8.2), which can also be called organizational memory. In this configuration, designers, manufacturers, certifiers and users can more easily share documents. In terms of the Orchestra Model, we could talk about music theory and style of music, as well as sets of scores for playing a symphony. However, documentation is still separated from the product.

If we push advanced interaction media much further and integrate it in the product, we can see it as a user interface (Fig. 8.3). Today, we have onboard information systems (OIS) in advanced commercial aircraft cockpits such as the A380. Of course, an OIS does not include all pieces of information, just the ones that are necessary for

Fig. 8.2 Overall technical documentation as a shared database

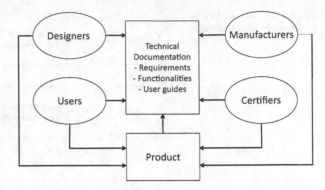

Fig. 8.3 Overall technical documentation as a user interface

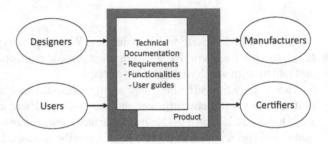

flying and managing systems. However, we are moving in the direction of electronic operational/technical documentation being connected to the product itself, which provides new contextual indexing capability.

This evolution can be counterproductive if it is not thought of as an integrated product from the beginning of the design process. Indeed, adding more useful functionalities can be great when they are taken individually, but this approach is likely to become a nightmare when people need to manage them altogether. Integration is a key issue that requires a well-orchestrated outside-in approach to design, which logically induces natural user interfaces in the end.

Finally, technical documentation is not only an output of design, it can be a very useful input to design. Indeed, design is an iterative process that requires both creativity and rationalization, as already mentioned in this book. In fact, the quality of technical documentation (of the design process and its solution) contributes to the quality of design itself. The reader of a document is the active part of the related reader-document system. In the same way and more generally, the user of a product is the active part of the user-product system. Continuing this documentation-product analogy, organization of a document (system), the way phrases (objects) are written (designed), as well as style and lexicon being used strongly influence reader's (user's) activity. The cognitive activity of the reader (user) should not be mobilized on interaction problems, and should be centered on understanding and interpretation of (active) document content instead.

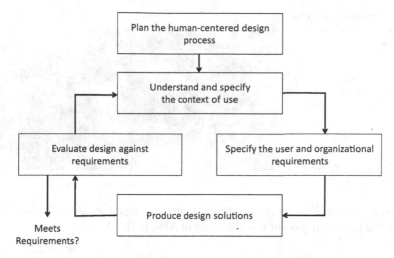

Fig. 8.4 Key human-centered design activities. (Adapted from ISO 13407)

As Norman (1992) already suggested, writing could be thought of as design and design as writing. We write for potential readers, as we design for potential users. We know that papers we write must be reviewed by several persons before being delivered outside, as we know that artifacts must be tested by several persons before being delivered outside. One of the most beautiful things in the evolution of hypertext technology is to observe that we can consider **hypermedia reading as human-computer interaction** (HCI), and **hypermedia writing as human-centered design** (HCD). This evolution converges toward both useful modeling and simulation tools, and integration of old operational documentation into the user interface. It also facilitates agile approaches to design and development. Finally, it enables us to better define high-level requirements, carry out flexible redesign using formative evaluations, and potentially support human-in-the-loop simulations. It is also necessary to develop principles and criteria for evaluation of new design solutions (Boy 2008). This approach is compatible with the human-centered design activities provided by ISO 13407[7] (Fig. 8.4).

The Virtual Camera as an AIM Example

Since the beginning of this book, I insisted on the need for human-in-the-loop simulations where prescribed cognitive functions can be tested and emerging cognitive functions discovered . M&S cannot be carried out without appropriate means, which we call Advanced Interaction Media supported by advanced HCI techniques and tools. AIM technology is evolving everyday, and what I will describe here is a schematic account of what we have done in a project, still in progress, that promotes

---

[7] http://www.iso.org/iso/catalogue_detail.htm?csnumber=21197.

Fig. 8.5 Example of use of a "third person view" for parking a car

the virtual camera (VC) concept currently being developed at NASA. This example will enable the illustration of possible uses of AIM in HCD.

Definition and Initial Set of Properties

The VC concept emerged from the early test of the Space Exploration Vehicle (SEV) developed by NASA for surface exploration of planetary bodies, including near-Earth objects, the Moon and Mars. It was previously called Lunar Electric Rover (LER) and Small Pressurized Rover (SPR). Indeed, driving a vehicle in a little-known environment is a difficult task. Even in a well-known environment such as reconstructed scenery of the moon at Johnson Space Center, we realized that the astronaut driving the SEV needed external advice to move safely. The idea of a virtual camera came up as a "third person view" (Fig. 8.5), as if someone outside the vehicle was able to see the scene and help the driver move safety and efficiently (Boy et al. 2010).

Such situations on the Earth are very common. However, it is barely possible on a partially-known planet to be explored (i.e., nobody will be available outside to guide the human operator). This is why we thought about using a readily available existing database (e.g., Google Moon) to provide increased **situational awareness**. This existing database includes already explored areas and unexplored regions. Astronauts driving an SEV (Fig. 8.6) will be able to use this existing database for navigation guidance and safety management.

The virtual camera uses such knowledge (i.e., existing database of the planet to be explored), and incrementally fuses in the database new data from physical sensors appropriately located (i.e., either on the rover or other vehicles also moving on the planet to be explored). We extended the VC concept to more general planetary exploration. In particular, when robots perform the exploration from Earth. The VC is being designed to support risk mitigation by providing the ability to investigate possible futures based on best possible data as well as choosing the right tools and systems to achieve the mission.

A VC is a software entity that requires four main entities (Fig. 8.7):

Fig. 8.6 LER on top of a hill (observed and represented 2D scenes)

Fig. 8.7 Virtual camera
high-level diagram

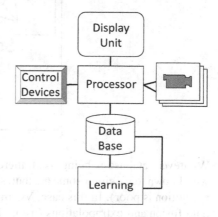

- A geographical/geological database (GGDB) that enables storing and using existing data about the surface of a planet or asteroid (e.g., information on the nature of the soil). This means that each pixel is n-dimensional (i.e., 3D and other dimensions relevant to geological knowledge). We will start with a well-mastered database such as Google Earth (or Google Moon) to demonstrate feasibility of the VC, and define the GGDB structure for further implementation.
- A user interface (display unit: DU) and manipulation mechanism (control devices: CD) that enable people to interact with the GGDB in the form of a virtual camera (e.g., VC used as a support for flying near an asteroid).
- A learning mechanism that enables integration of new geographical and/or geological data into the existing GGDB. Since we will incrementally find out about the content of such a database, its structure will need to be redefined incrementally. Consequently, we start with a well-mastered database (e.g., Google Earth), and test various properties of the VC on it in order to derive actual specifications of the final VC GGDB.
- A processor that synchronizes GGDB, DU, CD and the learning mechanism with external devices such as physical cameras, various kinds of sensors including mass spectrometry.

Fig. 8.8 Attitude parameters

Fig. 8.9 Virtual camera
generic display

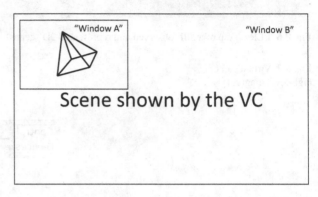

Whatever database is being used, there will be areas of the scene that are not very well known (e.g., either some attributes of the scene are not sufficiently known or the resolution is poor). In this case, VC may adapt the scene using data augmentation, data fusion and extrapolations (using the learning aspect of the VC). In addition, a virtual camera includes augmented reality features that either compensate 3D data-poor scenes or provide useful interpretations and advice to the user. VC requirements are the following: (x, y, z) position handling; (roll, pitch, yaw) attitude handling (Fig. 8.8); focus and zoom handling; a precision mode; (Vx, Vy, Vz) velocity.

The overall view of the VC with respect to location of the space vehicle will be displayed in a window "A", and the view provided by the VC will be displayed in a window "B" (Fig. 8.9). A simulated visualization is provided in Fig. 8.10.

Usability of these options still has to be tested (i.e., ease of learning, efficiency and precision, recovery from errors and subjective feeling in the manipulation). Note that VC control can be done either directly in the vehicle itself (relative reference) or from a remote station (absolute reference to be defined). In any case, user's perception of a VC tool has been observed to be a "smart window" on the world being explored.

It can be said that a virtual camera is a software object that is able to move in 3D space around virtual objects, a simulated rover or a robot for example, to provide the view of these objects in their environment from the point where the camera is located (Fig. 8.11). Obviously, a VC should be easy to manipulate and visualization should be clearly understandable and affordable. The VC control and display unit (CDU) should enable its user to get an appropriate mental representation of the

Fig. 8.10 Example of visualization using the virtual camera

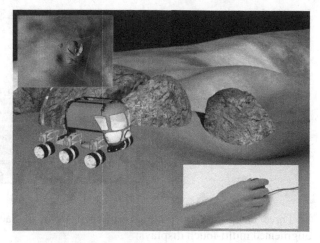

Fig. 8.11 An example of virtual camera use in an SEV

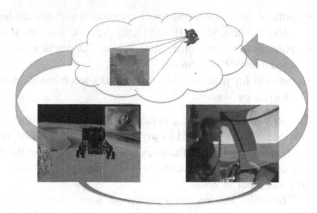

actual situation. Three types of CDU can be investigated as solutions: multi-touch and audio; joystick, and tablet.

Regarding multi-touch interaction patterns, here is a list of currently implemented patterns: Spread; Pinch; Drag; Flick; Tap; Hold; Spin.[8] In addition, to direct interaction using multi-touch screens, we can also use tablets to interact with those screens. This is one of the solutions that HCDi is trying to develop for nuclear power plant control rooms (Stephane 2012). An example of a solution is the interaction with large screens using iPods and Tablets PCs (Fig. 8.12).

The combination of both modalities, audio and multi-touch, tremendously enhances the quality of interaction both in precision and access time. A question is the directionality of sound input, but we have now systems that enables to direct sound inputs very accurately.[9]

---

[8] http://ui-patterns.com/blog/Exploring-new-patterns-multi-touch-displays.

[9] http://www.youtube.com/watch?v=bgz7Cx-qSFw.

Fig. 8.12 Interacting with a
large screen using a tablet PC
(HCDi work)

Three categories of patterns emerged from interactions with such audio-augmented multi-touch displays:

- patterns for augmenting touch interaction with speech-based interaction (Auditory Mode Switch, Voice-Based Distal Access, Voice as Text Input);
- patterns for augmenting speech-based interaction with touch interaction (Multimodal Error Correction);
- patterns for interweaving touch and speech-based interaction (Select by Touch, Operate by Voice).

The joystick solution is a better solution when a pilot uses the VC for navigation purposes. Pilots are used to joysticks, and this solution typically tends to decrease the learning time and efficiency during operations in the SEV. It enables 8 degrees of freedom (dof) on the same joystick or on two joysticks: 3 dof in position; 3 dof in attitude; 2 dof in zoom.

The tablet PC has been chosen as a solution that could be used in the spacecraft, (e.g., for the exploration of an asteroid) and on the ground (Platt and Boy 2012). It provides images in the relative frame of reference of the camera (i.e., the viewer directly sees the scene through the camera). From an AIM point of view, the tablet PC solution requires the following specifications. We can use screens such as Fig. 8.13 to help further define requirements for: Safety resolution requirements; Exploration resolution requirements; Real-time data display requirements; Format of data displays (visualization); Graphical User Interfaces.

This ongoing project is a good example of an HCD endeavor. We started from a series of brainstorming involving several experts, such as astronauts, space scientists and engineers, computer scientists, psychologists and artists. The VC concept emerged from this active HCD process as a necessity. We can say that it is a hard problem because it requires a multi-disciplinary approach.

More generally, useful and usable technology is developed by chance and necessity, as a living organism that exists because it has to be there. Coming back to the Orchestra Model, the VC is like a musical instrument that is required to play a specific symphony, in this case planetary exploration. Let me expand the story into a multi-agent world where VC information sources are distributed among a set

Fig. 8.13 Prototype screen
for VC showing areas of
interest and safety concern
with possible use of colors
and symbols

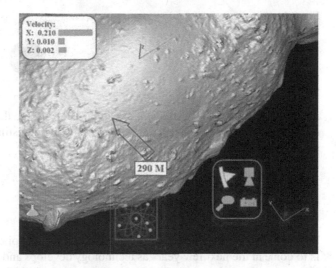

of agents. In planetary exploration, a set of robots can be used to get information from various appropriate viewpoints (e.g., from various locations, types of sensors). Data coming from these robots can be integrated into the VC database in real-time to improve both precision and nature of available information.

Indeed, the Virtual camera for space exploration purposes happened to be a more general concept than expected. Indeed, we are also applying it to weather forecasting in commercial aviation in another research project (Craig 2012). The lack of capacity associated with inappropriate weather management leads to delays that are more and more unpredictable (Donohue and Shaver 2008). Weather situational information is crucial in a pilot's decision making. For example, weather conditions that require immediate evasive action by the flight crew, such as isolated heavy rain, micro bursts, and atmospheric turbulence, require that the flight crew receive near real-time and precise information about the type, position, and intensity of those conditions. The VC concept appeared to be a good solution to display the various sources of weather information to pilots. In addition, the VC as forecasting support is used as a decision-making and planning tool, a kind of support that is urgently needed in air transportation, taking into account the constant increase in traffic density around big airports. Pushing the VC concept further, we could imagine that information coming from infrared sensors of other aircraft flying near a bad weather location be integrated into existing evolutionary databases. Such real-time data would be more accurate than what we have now, and pilots could easily navigate into the resulting database using the VC. We could either have a tablet or a joystick-based interaction. These solutions are still under investigation.

Therefore VC, as a general concept, can be formalized as follows (Fig. 8.14). Let D_{VC} be a database that includes scenes having a set of properties (P_1, P_2, ... P_n). Examples of such properties are color, shape and density. Let A_{VC} be a set of agents that can sense and provide data (d_{S1}, d_{S2}, ... d_{Sm}) that can be fused in D_{VC}. Such agents could be specialized robots or people. Data can be pictures, infrared data or

Fig. 8.14 Virtual camera as a
general concept

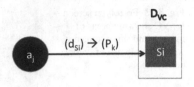

mass spectrometry data. An agent a_j can provide fusable data on scene Si enabling
modification of properties (P_k, k being variable). Interesting scenes can be named
and registered into a pattern database.

Conclusion

This chapter introduced important AIM concepts and applications. Of course, more
is to come in the next few years as technology develops and new user needs emerge.
Human-computer interaction continues to progress within the computer science field.
Since its inception, mainly with the birth of office automation, it was always associated with design, and human-centered design in particular. Today, its exposure is
much larger than office automation, and embraces all sectors of life (e.g., industrial
sectors; public sectors and homes). I prefer to call this new field of research and
practice, advanced interaction media, because the underlying technology mediates
interactions among agents whether they are humans or machines.

AIM can take several kinds of forms from paper documents to interactive technology. Newspapers support is currently evolving from paper to Internet. However,
while this evolution is limited to a change of support, it is also associated with a
change in semiotics. For example now, we can use animations smartly mixed with
text and pictures. We can hyperlink anything relevant. The way of "writing" is also
evolving.

AIM is a great technology that is providing support to different practices in design.
It enables rapid prototyping, and easy exchange of materials among design team
members. It enables the design team to create a chain of design solutions, which not
only materialize ideas, but also enable sharing these concrete ideas among involved
people. AIM brings proposed solutions closer to the real world than discussions and
paper could do before. Ultimately combined with modeling and simulation, it enables
human-in-the-loop simulations, and creates great environments for usefulness and
usability testing from early stages of the life cycle of a product. From that point of
view, AIM is a necessary element of modern systems engineering.

AIM will be integrated in a large number of products that will support our lives.
Since HCD deliberately takes the outside-in approach (i.e., the architect approach),
why don't we start by developing AIM from the beginning of a design project and
work on it until product decommissioning? AIM can be used for design, but it can
also support certification, training, operations, maintenance and the often forgotten
decommissioning. AIM will become a support for managing a product's life cycle.
It will provide visual, and more generally sensorial, situation awareness of what was

planned to be done, what was effectively done, why and how it was done, and so on. If we take the Orchestra metaphor, AIM will support the transfer of a symphony (i.e., a product) from composers to musicians and conductors (i.e., our human and machine multi-agent world), and ultimately to the audience (i.e., you!).

References

Abowd, G., & Mynatt, E. (2000). Charting past, present, and future research in ubiquitous computing. *ACM Transactions on Computer-Human Interaction, 7*(1), 29–58.

Berners-Lee, T. (1990). *Information management: A proposal.* This document was an attempt to persuade CERN management that a global hypertext system was in CERN's interests. Note (http://www.w3.org/History/1989/proposal.html) that the only name Tim Berners-Lee had for it at this time was "Mesh". He decided on "World Wide Web" when writing the code in 1990.

Berners-Lee, T., Hendler, J., & Lassila, O. (2001). The semantic web. *Scientific American Magazine.* May.

Boy, G. A. (1986). An expert system for fault diagnosis in orbital refueling operations. *AIAA 24th Aerospace Sciences Meeting*, Reno, Nevada, USA.

Boy, G. A. (1987). Operator assistant systems. *International Journal of Man-Machine Studies, 27,* 541–554. In G. Mancini, D. D. Woods, & E. Hollnagel (Ed.), *Cognitive engineering in dynamic worlds.* London: Academic.

Boy, G. A. (1991a). *Indexing hypertext documents in context. Proceedings of the ACM Hypertext'91 Conference.* San Antonio: ACM Digital Library.

Boy, G. A. (1991b). *Computer integrated documentation.* NASA-TM-103870. Ames Research Center, California, USA. This TM is synthesis of the work done during the 1989–1991 period at NASA Ames Research Center in the Artificial Intelligence Research Branch.

Boy, G. A. (1998). Cognitive function analysis for human-centered automation of safety-critical systems. In *Proceeding CHI'98, the ACM Conference on Human Factors in Computing Systems* (pp. 265–272). Los Angeles: ACM Digital Library.

Boy, G. A. (2002). Procedural interfaces. In *Proceeding IHM'02 (the Francophone Conference on Human-Computer Interaction)*, Poitiers, France. Also in New York: ACM Digital Library. (http://dl.acm.org).

Boy, G. A. (2005). Knowledge management for product maturity. *Proceedings of the International Conference on Knowledge Capture* (K-Cap'05). Banff, Canada. October. Also in New York: ACM Digital Library. (http://dl.acm.org).

Boy, G. A. (2008). *Human-centered development of perceived complexity criteria: Developed criteria.* Technical Report DGAC/EURISCO No. T-2007-201. Direction Générale de l'Aviation Civile, Bureau des Aeronefs et de l'Exploitation, Paris, France.

Boy, G.A., & Ferro, D. (2001). Using cognitive function analysis to prevent controlled flight into terrain. Chapter of the Human Factors and Flight Deck Design Book, Hashgate Pub., ISBN 0754613801.

Boy, G. A., Mazzone, R., & Conroy, M. (2010). The virtual camera concept: A third person view. *Third International Conference on Applied Human Factors and Engineering*, Miami, Florida; 17–20 July 2010.

Buxton, W. (1997). Living in augmented reality: Ubiquitous media and reactive environments. In K. Finn, A. Sellen & S. Wilber (Eds.), *Video mediated communication* (pp. 363–384). Hillsdale: Erlbaum. An earlier version of this chapter also appears in *Proceedings of Imagina '95*, 215–229. http://www.billbuxton.com/augmentedReality.html. Accessed 1 Sept 2012.

Card, S. K., English, W. K., & Burr, B. J. (1978). Evaluation of mouse, rate-controlled isometric joystick, step keys, and text keys for text selection on a CRT. *Ergonomics, 21,* 601–613.

Craig, C. (2012). Improving flight condition situational awareness through human centered design. *Symposium on Human-Centered Design of Life-Critical Systems, World Congress of Ergonomics (IEA 2012)*, Recife, Brazil.

Donohue, G., & Shaver, R. (2008). *Terminal chaos: Why U.S. air travel is broken and how to fix it.* Reston: American Institute of Aeronautics and Astronautics, Inc.

Dourish, P. (2001). *Where the action is: The foundations of embodied interaction.* Cambridge: MIT. ISBN 0262041960.

Engelbart, D. C. (1966). *Study for the development of human intellect augmentation techniques.* Quarterly Technical Letter Report 1. May. Stanford University. http://sloan.stanford.edu/mousesite/EngelbartPapers/B2_F5_ARNAS1.html. Accessed 1 Sept 2012.

Epps, B. W. (1986). Comparison of six cursor control devices based on Fitt's law models. In *Proceedings of the Human Factors Society 30th Annual Meeting* (pp. 327–331). Santa Monica.

Fitts, P. M. (1954). The information capacity of the human motor system in controlling the amplitude of movement. *Journal of Experimental Psychology, 47*(6), 381–391.

Grudin, J. (1994). Computer-supported cooperative work: History and focus. *Computer, 27*(5), 19–26.

Heidegger, M. (1962). *Being and time.* Originally published in 1927. New York: Harper and Row.

Langeheinrich, M. (2001). Privacy by design—Principles of privacy aware ubiquitous systems (pp 273–291). In UBICOMP 2001, LNCS 2201.

Leiner, B. M., Cerf, V. G., Clark, D. D., Kahn, R. E., Kleinrock, L., Lynch, D. C., Postel, J., Roberts, L. G., and Wolff, S. S. (1997). The past and future history of the Internet. *Communications of the ACM, 40*(2), 102–108.

MacKenzie, I. (1992). Fitts' Law as a research and design tool in human computer interaction. *Human Computer Interaction, 7,* 91–139.

Marill, T., & Roberts, L. G. (1966). Toward a cooperative network of time-shared computers. Fall AFIPS Conference, October. http://www.packet.cc/files/toward-coop-net.html. Accessed 23 Jan 2012.

Mindell, D. A. (2008). *Digital Apollo—Human and machine in spaceflight.* Cambridge: MIT. ISBN 10:âŁ˙0-262-13497-7.

Nielsen, J. (1993). *Usability engineering.* Boston: Academic. ISBN 0-12-518405-0.

Norman, D. A. (1992). *Turn signals are the facial expressions of automobiles.* Cambridge: Perseus.

Norman, D. A. (1998). *The invisible computer: Why good products can fail, the personal computer is so complex, and information appliances are the solution.* Cambridge: MIT. ISBN âŁ˙0-262-14065-9.

Platt, D. W., & Boy, G. A. (2012). The development of a virtual camera system for astronaut-rover planetary exploration. *Symposium on human-centered design of life-critical systems, World Congress of Ergonomics (IEA 2012)*, Recife, Brazil.

Saaksvuori, A. (2008). *Product lifecycle management.* Heidelberg: Springer. ISBN 3540781730.

Schultz, A. (1973). *On phenomenology and social relations.* In H. Wagner (Ed.). Chicago: University of Chicago Press.

Stark, J. (2004). *Product lifecycle management: 21st century paradigm for product realization.* Berlin: Springer. ISBN 1-85233-810-5.

Stark, J. (2007). *Global product: Strategy, product lifecycle management and the billion customer question.* Berlin: Springer. ISBN 1-84628-914-9.

Stephane, A. L. (2012). Advanced interaction media in nuclear power plant control rooms. *Symposium on Human-centered design of life-critical systems, World Congress of Ergonomics (IEA 2012)*, Recife, Brazil.

Van Dam, A., & Rice, D. E. (1970). Computers and publishing: Writing, editing and printing. In *Advances in computers* (Vol. 10, pp. 145–174). London: Academic.

Ware, C., & Mikaelian, H. H. (1987). An evaluation of an eye tracker as a device for computer input. In: Graphics Interface '87 (CHI+GI '87), (pp. 183–188 April).

Chapter 9
Conclusion

HCD is not about human factors and ergonomics as a discipline that is used when systems are already designed and developed to fix cosmetic design flaws. HCD is not a refinement of human-computer interface design, techniques and tools. HCD is about reinventing engineering and design into a single discipline that integrates technology, organization and people. HCD is a discipline that enables the generation of more appropriate high-level requirements that lead to successful products. HCD recognizes that in any innovation there will be emergent properties that need to be identified. HCD supports incremental refinement of the **integration of technology, organization and people** (iTOP) towards maturity. HCD of life-critical systems is necessarily based on human-in-the-loop simulations in order to understand intimate relationships between technology, organization and people, and identify emergent properties.

HCD cannot be implemented without the right people in the right place. HCD requires socio-technical leadership at all levels of the organization. HCD requires competence, talent and engagement. HCD is certainly the renaissance of engineering for people's safety, efficiency and comfort. HCD cannot be thought of without a clear understanding of the life cycle of the product being designed, including its obsolescence and decommissioning.

This **renaissance of engineering** cannot be developed without breaking the current finance-driven dictatorship. Instead of having two dependant variables, costs and profits, that are optimized to the sacrifice of "independent" variables such as human-centered quality, usability and usefulness, I make a distinction between human-centered quality that is supported by educated common sense testing from currently implemented quality management that is mechanical and involves very little of human expertise and experience. For that matter, I share the view of Bob Lutz, a former vice-chairman of General Motors, who recently stated:

> It's time to stop the dominance of the number crunchers, living in their perfect, predictable, financially projected world (who fail, time and again), and give the reins to the "product guys" (of either gender), those with vision and passion for the customers and their product or service. (Lutz 2011).

G. A. Boy, *Orchestrating Human-Centered Design,*
DOI 10.1007/978-1-4471-4339-0_9, © Springer-Verlag London 2013

From Linear/Local to Non-Linear/Global

At a deeper level, it is also important to acknowledge that, during the twentieth century, engineering was mainly based on local approximations, and mainly supported
by linear algebra. The complexity of such systems forced engineers to **simplify**
them in **specific contexts**. Linearization is generally very good within the limited
context where it is applied (i.e., nominal situations), but it can considerably rigidify
operations in other situations, such as off-nominal, emergency or unexpected situations where flexibility is required. Automation (i.e., automation of technology) and
operational procedures (i.e., automation of people) are specific examples of such
linearization. Even if flexibility and automation rigidity are antagonist concepts, we
should find a framework that enables their association for the welfare of LCSs where
they are at stake.

Despite appearances, it is generally not natural complexity that disturbs people; it
can be this forced simplification, or linearization, that tends to remove human natural
defenses. Because it works in nominal situations most of the time, people tend to
become complacent and lose their immunological cognitive functions. This is one
of the reasons why, during the last 3 decades, human factors specialists focused on
human errors, and produced voluminous literature on human reliability at work using
modern technology. Even if the concept of resilience emerged in the human factors
literature, no real valid proposal was provided regarding the way human-centered
design should be done. I strongly believe that many errors are good to make because
we learn from them (Boy 1996). Successful design often resulted from engineering
failures (Petroski 1992). We are back to the issue of maturity.

Bainbridge (1983) discussed ways in which automation of industrial processes
may expand rather than eliminate problems with the human operator. This observation and analysis was done at the beginning of what automation is today (i.e.,
30 years ago). We did not yet rationalize the concepts of technological maturity and
maturity of practice. We also did not understand that we were moving into a more
complex world where the accumulation of linear approximations would create a gigantic network behaving as a complex system. She said that automation cannot leave
the operator with responsibility for abnormal conditions, and, guess what? This is
what we have done. We automated on the hat of the Gauss curve, leaving the residual
unanticipated parts to human operators. Therefore, crisis management has become a
real issue today.

Bainbridge talked about human-computer collaboration. This is precisely what
we are doing today. Consequently, we need to understand what such collaboration
means. In aviation, we start to better understand that not only pilots need to keep
their flying capabilities, but also learn how to handle automated systems and not take
them for granted. Humans and automated machines have to work in teams. We then
need to analyze novelty complexity and derive what the game rules are. **Functional
awareness** (e.g., what is the right level of granularity that should be provided in
order to ensure a sufficient situation awareness?) is a main issue in the control and
management of a complex system. Functional awareness is a matter of context. In

some cases, one level of granularity is useful and appropriate. In other cases, another level of granularity can be more useful and appropriate. Again, the use of the AUTOS pyramid is always helpful in structuring solutions compatible with this functional awareness problem.

We need to focus more on what complexity means instead of continuing to linearize our lives piece by piece, no matter what! The irony today is that the main beneficiaries of linearization are accountants and lawyers. Why? This is because linear systems are more measurable. In addition, when you have only one variable unit, that is money, it is easier to develop powerful software-based systems and master them.

Human beings are unique because they adapt. Adaptation is a constant loop process. People adapt to complex systems by developing skills and knowledge. They need to maintain active stability, and therefore are necessarily aware of what is going on... otherwise they die! Life and death are dual concepts. Today, we almost assume that safety is insured by linearized structures build by humans. When this insurance fails, people are lost. This is why we need to restructure our comprehension of risk taking (ASA 2008).

It is time to rehabilitate a humanistic view of engineering that acknowledges people as auto-organized structures, coupled through recursive processes of self-regulation, and ultimately governed by their deeper intentions. Human-machine systems are **complex** because of such auto-organization and self-regulation. When life is at stake, it is preferable to use the term **life-critical system** instead of "human-machine system", knowing that an LCS includes both people and machines.

Technology and organizations get old in the same way as people get old. We tend to forget some lessons learned over time. Complacency progressively replaces reinforced situation awareness, articulation, coordination and involvement. Both organizations and people can be modeled as societies of agents (Minsky 1985), which need to perceive, understand and anticipate both their internal and external interactions. These interactions can be managed by supervision, mediation or cooperation by mutual understanding (e.g., refer to the chapter on the Orchestra Model in this book). Articulation work is crucial to keep appropriate activity and results within an organization of agents (i.e., some agents articulate workflows among the other agents), whether they are people or organizations. Coordination among agents is also mandatory when they need to produce through a collaborative activity. Finally, agents are required to be involved to produce acceptable work results.

Standardization of Human Factors Versus Human-Centered Design

Human reliability in life-critical systems has been one of the major human factors research topics for the last 3 decades (Swain and Guttman 1983; Norman 1988; Reason 1990; Hollnagel 1991; Leveson 1995; Hollnagel et al. 2006; Dismukes et al. 2007). Parts of these research efforts led, and still lead, to the development of regulations such as ISO 14001 and EASA CS 25.1302. Today, commercial aircraft are typically certified using such regulations. More regulations are in preparation by national and international regulatory authorities.

This increasingly normative approach attempts to incorporate many human-like abilities directly into design and development processes. If it reflects recognition of the human factors discipline, we could argue that deeper principles of that discipline are not necessarily being implemented and carried out. There are strong similarities with quality assurance practice in industry, where process quality is carefully certified but product quality remains an issue.

We tend to automate organizations in the way we automate technology. Consequently, we start to observe similar symptoms such as surprises (Sarter et al. 1997). As seen previously, both procedures and technological automation rigidify work, and therefore eliminate flexibility when it is necessary. In addition, such approaches to engineering and production in large industrial organizations tend to remove appropriate human expert involvement in everyday problem solving and assessment of the product being developed. Human experts are mostly used in crisis management. This new practice is developed because financially-driven organizations believe that they can reduce costs by proceduralizing, in the form of reporting and regular activity planning, and automating. In nominal situations, expert people are replaced by cost-effective personnel who fill forms instead of technically think, do and refine. The problem is that situations are not always "nominal". Again, syntax has become more important than semantics. This kind of evolution would be fine if the overall system was mature. Unfortunately, it is not.

HCD tries to define not only technology maturity (e.g., in terms of product availability), but also maturity of practice (e.g., in terms of successful product appropriation with respect to safety, efficiency and comfort) and now organizational maturity (i.e., in terms of information and workflow, people motivation and involvement, and socio-technical stability). The Orchestra Model is important to consider because people at the bottom of the chain have moved into it (i.e., musicians know how to play together), but organizations are still managed as armies using financial constraints and reporting as music theory. This mismatch between the two models, army and orchestra, creates problems that are difficult to solve because of inconsistencies of the various expectations. We do not have appropriate chief engineers (i.e., composers) who are capable of orchestrating the various jobs for the sake of an integrated product in the end (i.e., the symphony). Finally, we do not have a real leadership at the top of large companies (i.e., conductors), which is capable of managing the actors (i.e., the musicians) for the actual development of the product.

An Opportunity for STEM Education

It is time to rehabilitate science, technology, engineering and mathematics (STEM) and offer motivating programs to young people. I strongly believe that HCD is the future of engineering, and needs to be fed by not only science and mathematics but also art. In other words, STEM must be **making STEAM** (!), where the "A" means Art. We already discussed the difference between engineering and design. Engineering makes products using an inside-out approach, while design makes products using an outside-in approach. We need to generate a new breed of architects who not only know about engineering, but also know how to make esthetic, useful and usable things.

Cognitive engineering opened the way by promoting the view that information technology introduced a shift from doing to thinking (Norman 1986; Hollnagel and Woods 2005). Today, we are on a different trip; we are going **from thinking to being** (Latour 1986, 2011). This semiotic approach has been investigated for a long time, but we are just entering into its realization. Technology enables us to do things that were not possible before. For example, distance is no longer a real issue; we can communicate with almost anyone on the planet; we can even go and see someone or something when we decide to do it. However, it is important to realize that it is because technology enables this kind of possibilities that are who we should be today (i.e., goal-driven and can make choices), instead of be prisoners of technology that constrains us to do things (i.e., event-driven and dependent on technology). This is a question of maturity of practice.

As a reminder, semiotics is divided into three disciplines: syntax (i.e., relations among signs (or terms); syntax is about formal structures); semantics (i.e., relations between signs and concepts or objects that signs denote; semantics is about meaning); and pragmatics (i.e., relations between signs and the effects they have on the people who use them; pragmatics is about affordances). A large amount of work has been done in syntax and semantics in human-computer interaction. Pragmatics is more recent; in 1977, Gibson coined the term "affordances " to denote the capacity that an object has to suggest an action from its user (Gibson 1977). Today, information technology capitalizes various kinds of interactive objects that include appropriate syntax, semantics and pragmatics.

It is time to use information technology (IT) to support STEAM education. STEAM is about innovation and design, as well as, more importantly, knowledge design. Using IT, students can build their own abstract world as well as concrete innovative products. They obviously will need to study how to use science and mathematics to make innovative things in their own "laboratory" (Edwards 2010). It takes creativity and leadership to change the world we live in. Therefore, Art should be part of the trip. Edwards designed and conduct his *Lab* at Harvard University as a cultural incubator for education. The claim is simple: it is by taking risk and making innovative things that people learn. A solution is to create laboratories where artists and scientists collaborate to produce intriguing cultural content and surprising innovations.

We have seen in this book that modeling and simulation, associated with advanced interaction media, enables great support to human-centered design. Now, M&S and AIM, associated with art, are likely to improve creativity and innovation. The very fact that M&S and AIM are being used almost guaranties the rationalization of the product (i.e., the rational management of its life cycle). I claim that incremental operational tests and refinement of the product being designed typically lead to maturity, even when the product is STEAM education.

Entering into a Complexity Culture

If the twentieth century was the revelation of Descartes's theories, I strongly believe that the twenty-first century is going to be the revelation of Leonardo da Vinci's creative thinking. Again, we are moving from linear/local to non-linear/global. We are moving into a human and machine multi-agent world that requires management capabilities, involving situation awareness, decision-making and planning. We are moving from manual control to cognitive management. Complexity is around us as a web of interconnections (e.g., the World Wide Web and social networks). Creativity helps managers to capitalize on complexity. We need to invent solutions, test them, and re-test them until they are satisfactory. This is the only way we can reach maturity of both solutions and ourselves.

Creativity is not only about thinking, it is also and foremost about producing things and testing how they could be useful and usable. When you have an idea, you have to make it concrete. To do this you need to create the means to achieve results. It is not always easy, but rewards are up to the efforts. You need to have the courage to create: "Creativity is the process of bringing something new into being... creativity requires passion and commitment" (May 1994).

Even if we need to continue capitalizing on our past discoveries and knowledge, education cannot only look backwards, it has to look toward the future. Educating and training the next generation of leaders in science and technology requires design thinking and new kinds of conceptual backgrounds and tools, such as complexity science. We now need to address global issues that have to be solved using new types of science and technology. Teamwork is one of the components of this endeavor. We cannot forget that this will not be achieved without vision and passion. This is why our lab is the world. Our globalized economy is not working today because there is no governance, no global order. Current economical freedom leads to financial anarchy, and obviously benefits to a few people, but handicap the rest of us. We need to educate the future leaders, who are scientists, engineers, entrepreneurs, and policymakers. This is a worldwide enterprise. Without interdisciplinary communication on global issues, there will not be long-term coordination.

The question is: can we teach complexity science at school knowing that mathematics and science are currently very badly absorbed by children? It seems that a solution is to change our way of thinking regarding success and failure, and project positive attitude toward children and young people. I recently worked with my graduate students at FIT and the Ecole Polytechnique in Paris on why young people do not chose STEM education and consequently STEM jobs. All students involved in this study had excellent STEM knowledge and skills. They all said that there is a huge gap between what is taught at school and what is required in engineering schools and more generally in industry. They all love STEM but did not know how to apply their knowledge and skills in their future professional life. Many of them chose to go to business schools and get an MBA because they not only could get more money but also they could use their STEM knowledge and skills more effectively in doing business instead of engineering and science. The question is then, if the best people are going to business what will they sell in the long run? It is time to refocus on education and reformulate what we really need to learn.

Creativity Must be Associated with Rigorous Methods

I introduced Art and complexity in this conclusion as an unavoidable necessity in the early twenty-first century. However, it is crucial to keep in mind that rigor and concentration are also necessary (i.e., the right brain is important, but the left brain is important also). Consequently, we need to develop methods and tools that will support this new endeavor... to make reliable and efficient STEAM.

We have seen that technology often results in complacency. People tend to rely on technology especially when it works all the time. However, it sometimes fails, and people need to fix things. They are not always prepared to do this. There is still a long way to go to understand the core of human and machine symbiosis. What is the right function allocation among human and machine agents? Since Fitt's law, we did not make much progress (Fitts 1951). We now need to investigate this kind of hard problems within a different framework involving the Orchestra Model and complexity science. There is room for creativity of course, but without concentration and rigor nothing will be found, because we are facing very difficult issues.

We need to see complex systems as our strengths, instead of avoiding their investigation and coming back to inappropriate linear solutions. The Web is a great resource when it is well used. For example, problem solving is not limited to decomposing a problem into sub-problems until we find a small problem easy to solve. Problem solving can be done by a group of people well organized, and where skills and knowledge are distributed. Communication, cooperation and coordination are key processes. This is not new. Not so long ago, people need to organize themselves to solve complex problems in the countryside; for example, the whole village was involved in harvesting. Communication, cooperation and coordination were assets to solve the harvesting problem on a recurrent manner. New communities emerge in social networks today to solve problems, exactly in the same way as for harvesting in the countryside village. Specific rules have to be followed to reach acceptable results.

Ethics and Empathy

On 28 January 1986, the Space Shuttle Challenger (STS-51 L mission) disintegrated in flight 73 s after take-off. An O-ring seal on a solid rocket booster failed at launch. At 59 s after launch, hot gasses pierced the shell of the booster, burning into the Shuttle's external hydrogen tank. This was a disaster due to a very small initial condition applied on a very complex system. When we talk about system here, we do not talk about the space shuttle itself only, but also about the gigantic complex organization that supported it.

Some people knew about O-rings vulnerability and catastrophic consequences before the flight. Some of them even tried to stop the flight, but Group Think phenomena prevailed. Morton Thiokol managers agreed that the issue was serious enough to recommend delaying the flight, but they finally complied with NASA decision, and made no objections to launching the Challenger space shuttle.

In particular, Roger Boisjoly, a Morton Thiokol engineer, wrote a memo to his management regarding the faulty design of the Space Shuttle's solid rocket boosters. He insisted on the fact that this flaw could lead to an accident during launch. Roger Boisjoly did not make the news. For me, he was the man to listen to. He was the expert. He was the first violinist. Nobody listened to him.

The President of the United States ordered a presidential commission to investigate the accident. Boisjoly was called as a witness, and gave his opinion on how and why the O-rings failed. Unfortunately, nobody took into account what he said. Consequently, he resigned from Morton Thiokol. In 1988, he was awarded the AAAS Prize for Scientific Freedom and Responsibility for his integrity leading up to and following the Challenger disaster.

In the future, it is important to revisit this kind of accident with ethics and empathy in mind. We need to recognize technical competence in decision-making. In complex systems, situation awareness is a matter of multi-agent mutual respect and competence recognition. Who do we trust? Decision-making is a matter of accumulating many mini-decisions and viewpoints, in order to make an overall impression of the situation, a sort of attractor from a complexity science standpoint. Real leaders must be able to do this.

It is not the point that I am convinced by theoretical arguments. In neither economy, nor politics nor strategy, there is not, I think, absolute truth. But there are circumstances. It is the idea that I make of the situation that makes me decide. (De Gaulle 1959)

References

ASA (2008). *Academy of air and space conference on risk-taking electronic proceedings.* http://www.academie-air-espace.com/publi/detail.php?&varID=150varCat=1. Accessed 12 April 2012.

Bainbridge, L. (1983). Ironies of Automation. *Automatica, 19*(6), 775–779.

Boy, G.A. (1996). Learning evolution and software agents emergence. Proceedings of ITS96, Lecture Notes in Computer Science Series, Springer Verlag, Berlin, ISBN:3-540-61327-7, pp. 10–25.

de Gaulle, C. (1959). *Mémoires de guerre—Le Salut: 1944–1946* (tome III). Plon, Paris. Also in. Pocket, 1999 (new edition in 2006). ISBN 2-266-16750-2 & 978-2266167505.

Dismukes, R. K., Berman, B. A., & Loukopoulos, L. D. (2007). *The limits of expertise: Rethinking pilot error and the causes of airline accidents.* Aldershot: Ashgate.

Edwards, D. (2010). *The lab: Creativity and culture.* Cambridge: Harvard.

Fitts, P. M. (1951). Some basic questions in designing an air-navigation and air-traffic control system. In N. Moray (Ed.), *Ergonomics major writings* (Vol. 4, pp. 367–383). London: Taylor & Francis.

Gibson, J. J. (1977). The theory of affordances. In R. Shaw & J. Bransford (Eds.), *Perceiving, acting, and knowing.* ISBN 0-470-99014-7.

Hollnagel, E. (1991). The phenotype of erroneous actions: Implications for HCI design. In G. W. R. Weir & J. L. Alty (Eds.), *Human-computer interaction and complex systems*. London: Academic.

Hollnagel, E., & Woods, D. D. (2005). *Joint cognitive systems: Foundations of cognitive systems engineering*. Boca Raton: CRC/Taylor & Francis.

Hollnagel, E., Woods, D. D. & Leveson, N. (Eds.). (2006). *Resilience engineering: Concepts and precepts*. Aldershot: Ashgate. ISBN 0754646416.

Latour, B. (1986). Visualisation and cognition: Drawing things together. In H. Kuklick (Ed.), *Knowledge and society studies in the sociology of culture past and present* (Vol. 6, pp. 1–40). Jai. Reprinting and revision in M. Lynch & S. Woolgar (Eds.), *Representation in scientific activity* (pp. 19–68, 1990). Cambridge: MIT. Partial republication in M. Dodge, R. Kitchin & C. Perkins (Eds.), *The map reader. Theories of mapping practice and cartographic representation* (pp. 65–73, 2011). Wiley, Blackwell. Republication in M. Lynch (Ed.), *Science and technology studies: Critical concepts* (2011). Routledge.

Leveson, N. (1995). *Safeware: System safety and computers*. New York: Addison-Wesley.

Lutz, B. (2011). *Car guys vs. bean counters*. New York: Portfolio/Penguin.

May, R. (1994). *The courage to create*. New York: W.W. Norton & Company. ISBN 0393311066.

Minsky, M. (1985). *The society of mind*. New York: Simon and Schuster.

Norman, D. A. (1986). Cognitive engineering. In: D. A. Norman & S. W. Draper (Eds.), *User centered system design: New perspectives on human-computer interaction* (pp. 31–61). Mahwah: Lawrence Erlbaum Associates.

Norman, D. (1988). *The psychology of everyday things*. New York: Basic Books.

Petroski, H. (1992). *To engineer is human* (1st ed.). New York: Vintage Books, Random House, Inc.

Reason, J. (1990). *Human error*. Cambridge: Cambridge University Press.

Sarter, N. B., Woods, D. D. & Billings, C. E. (1997). Automation surprises. In G. Salvendy (Ed.), *Handbook of human factors & ergonomics* (2nd ed.). Wiley.

Swain, A. D., & Guttman, H. E. (1983). *Handbook of human reliability analysis with emphasis on nuclear power plant applications*. NUREG/CR-1278, Washington D.C.

Index

A

Abduction
  abductive inference, 51
Abstraction, 30, 130, 146, 154, 180
Accident, 2, 5, 13, 18, 23, 36, 54, 60, 66,
  72–76, 83, 89, 110, 117, 126, 140
Accountability, 10, 13, 19, 113, 126, 136,
  149, 180
Action reversibility, 104, 154
Active design document, 182
Activity
  activity theory, 30
Adaptation, 15, 26, 54, 66, 73, 80, 97, 102,
  125, 147, 152, 154, 157, 158, 163, 199
Aerospace engineering
  aerospace engineer, 1
Affordance, 47, 67, 109, 153, 161, 181, 201
Agent
  artificial agent, 98–100, 107
  human and machine agents, 9, 12, 14, 23,
    31, 38, 83, 143, 168, 203
  natural agent, 97, 98
Air traffic management (ATM)
  air traffic control (ATC), 10, 22, 23, 28, 30
Anticipation, 5, 36, 73, 159
Architect, 25, 32, 40, 41, 110, 111, 184, 201
Artificial intelligence, 17, 26, 38, 41, 48, 60,
  97, 151
Attractor
  cognitive attractor, 62, 155
Authority
  authority sharing, 14, 19, 22, 23, 25, 27, 28,
    80, 108, 120
Automation
  automation design, 10, 11
  automation of people, 38, 74, 198
  human-centered automation, 11
  organizational automation, 2, 10, 12–14,
    105, 120, 122, 123, 126–128, 133, 136
Autopoeisis
  autopoietic system, 118
AUTOS, 45, 73, 141–143, 146, 158, 163, 199
Availability, 21, 51, 64, 79, 102, 103, 125,
  157, 175

B

Black Swan, 79, 149
Black-box, 101

C

Catastrophe
  catastrophic properties, 5
Chronology, 18
Cockpit, 37, 38, 42, 46, 64, 66, 70, 107, 131,
  140, 146, 151, 156, 158, 181, 185
Cognitive function, 9, 11, 16, 23–26, 28, 98,
  121, 130, 134, 136, 141, 143, 149, 159
Cognitive psychology, 17, 38, 60
Collaboration, 27, 32, 184, 198
Comfort, 6, 25, 36, 42, 50, 59, 60, 67, 70, 90,
  100, 108, 110, 127, 131, 133, 161, 197
Common frame of reference (CFR), 9, 18, 20,
  24, 28, 126, 135, 167
Communication
  media, 5, 109
Competence, 21, 22, 36, 39, 82, 83, 105, 117,
  125, 130, 131, 135, 175, 204
Competition, 21, 30, 32
Complacency, 36, 63, 117, 128, 199, 203
Complexity
  artificial complexity, 4, 94, 95
  cognitive complexity, 153, 154
  complex system, 31, 38, 47, 74, 75, 89
  complexity analysis, 102, 146, 157
  complexity science, 5, 80, 147, 202, 203
  computational complexity, 153